Gasdynamic Theory of Detonation

Combustion Science and Technology

Editor

Irvin Glassman, Guggenheim Laboratories, Princeton University

VOLUME 1 Gasdynamic Theory of Detonation
Heinz D. Gruschka and *Franz Wecken*

Gasdynamic Theory of Detonation

Heinz D. Gruschka
The University of Tennessee
Space Institute

Franz Wecken
German–French Research Institute
Saint-Louis

GORDON AND BREACH SCIENCE PUBLISHERS
New York Paris London

Gordon and Breach, Science Publishers, Inc.
440 Park Avenue South
New York, N.Y. 10016

Editorial office for the United Kingdom
Gordon and Breach, Science Publishers Ltd.
12 Bloomsbury Way
London W. C. 1

Editorial office for France
Gordon & Breach
7–9 rue Emile Dubois
Paris 14e

Library of Congress catalog card number 70-1360502. ISBN 0 677 033702.
Printed in east Germany.

To Elsie and Lucia

Preface

Besides the well known textbooks about the detonation of solid explosives by Cook (1958) and by Berger and Viard (1962) there exist several texts which give a predominantly theoretical gasdynamic discussion of detonation, such as the article in *Fundamentals of Gasdynamics* (1958) and the book by Zeldovich and Kompaneets (1955, English 1960). Since that time, however, the detonation theory and in general the thermodynamics and gasdynamics of thermo–chemical nonequilibrium flows have been improved considerably. This is manifested in contributions to symposia and in numerous scattered publications. A typical aspect of modern investigations is, e.g., the distinction between frozen and equilibrium velocity of sound. A new presentation of the detonation theory which attempts to take these developments into account seemed to be justified.

This text has been developed out of a seminar which was held in 1966–67 at the German–French Research Institute Saint Louis (ISL), St. Louis, France. Certain preferences and limitations indeed result from this. The emphasis of the discussion is on solid explosives. Phenomena which are characteristic of gas detonations, like transition from deflagration to detonation, instabilities, and multidimensional structures, are therefore omitted. These phenomena are discussed in recent articles and monographs particularly by Russian authors. Instead, the interaction with inert surrounding materials has been regarded.

We gratefully acknowledge valuable and clearifying discussions with several of our colleagues, particularly with Dr. H. Behrens and Dr. G. Thomer. Our thanks are due also to the editor Prof. I. Glassman and to the publisher Gordon and Breach.

H. D. Gruschka
F. Wecken

Contents

CHAPTER I

Shock Waves

1 The Fundamental Gasdynamic Equations for Steady Shock Waves

IN ORDER TO DESCRIBE a shock wave in a moving gas we may restrict our discussion first to the one-dimensional steady plane case which already yields most of the basic results.

A general surface of discontinuity separating continuous flow regions may always be approximated *locally* by a stream tube flow, and may then be transformed to a steady, normal shock wave by application of the proper Galilei-transformations*.

1.1 *Conservation theorems*

We consider the following one-dimensional plane flow system consisting of three different regions, Fig. 1.1.

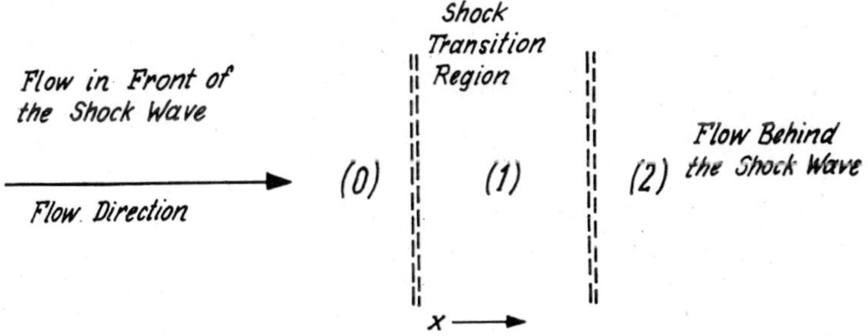

Fig. 1.1 One-dimensional plane flow through a normal shock wave.

Region (0) contains the gas in front of the shock wave (constant initial state).

Region (1) contains the gas within the shock transition region (variable state, dependent on x).

Region (2) contains the gas which has passed the shock (constant final state of the gas).

---

* See e.g. [2], p. 102.

The gas within these three regions has in general different states. The physical quantities describing the gas are labeled by subscripts 0, 1, 2 according to the three different flow regions considered.

We assume that our flow system is ruled by the basic conservation theorems of mass, momentum, and energy.

This system is not a closed system in the thermodynamic sense and cannot be transformed into a closed system in a simple way. Therefore, the conservation theorems take the form of balance equations for the three quantities. The fluxes in and out of the different regions bounded by control surfaces are equal.

The resulting conditions applied to the volume element of the flow yield the basic equations for the motion of the fluid. These equations—the continuity equation, the Navier–Stokes equation of motion, and the energy equation—integrated once for plane one-dimensional flow take the form of balance equations. The general integral form* of these conservation theorems would even be valid for flow regions including a finite number of discontinuities, e.g. shock waves.

The balance equations for steady, one-dimensional plane flow with heat conduction and viscosity are†:

Mass Balance (continuity equation)

$$\varrho u = \text{const} = \varrho_0 u_0 = \theta . \tag{1.1}$$

Momentum Balance (equation of motion)

$$\varrho u^2 + p - \hat{\eta}\frac{du}{dx} = \text{const} = \varrho_0 u_0^2 + p_0 = p_r . \tag{1.2}$$

Energy Balance (energy equation)

$$H + \frac{u^2}{2} - \frac{\hat{\eta}}{\varrho}\frac{du}{dx} - \frac{\lambda}{\varrho u}\frac{dT}{dx} = \text{const} = H_0 + \frac{u_0^2}{2} = H_r . \tag{1.3}$$

These equations contain the unknown quantities:

p, ϱ, u pressure, density, and x-component of the flow velocity, and
H, T specific enthalpy and absolute temperature.
θ the mass flux density, and

* For regions with continuously changing flow variables, in general the differential form of the basic equations is used; see e.g. [2].

† For the exact derivation reference is made to fundamental texts on gasdynamics, particularly [3], [5], [7], [6].

p_r, H_r the zero-velocity values of the pressure* and specific enthalpy;
$\hat{\eta}$, λ coefficients describing the influence of viscosity and heat conduction.

The specific enthalpy H and the absolute temperature T are specified as funtions of p and ϱ by thermodynamic equations of state.

The flow through a normal shock wave is completely described by Equations (1.1) to (1.3) if the initial state (p_0, ϱ_0, u_0) of the gas or fluid is given and the proper equations of state are known.

1.2 *Discussion of the fundamental equations*

The balance equations (1.1) to (1.3) together with the equations of state represent a system of five equations for the calculation of the unknowns p, ϱ, u, T and H.

The equations are valid provided that the variables are functions of the space coordinate x only. That is, the flow is one-dimensional and steady.

The physical meaning of the different equations and quantities is the following:

The *continuity equation* (1.1) states that if steady state conditions are assumed the *mass flux density* ϱu† has the same value everywhere. Neither spatial nor temporal fluctuations occur. The flow is steady and free of sources and sinks.

The *momentum balance*, equation (1.2), derived from the equation of motion, states that the *momentum flux density* ϱu^2 together with the forces acting in the fluid are invariable quantities. In other words: a decrease of the momentum flux density due to shock transition is compensated for by a pressure rise which must be overcome by the flow. If viscous forces are present, part of the pressure is due to this viscous resistance:

$$\varrho_0 u_0^2 - \varrho u^2 = p - p_0 - \hat{\eta}\,\frac{du}{dx} = \Delta\sigma_{xx},$$

σ_{xx} = diagonal element of the stress tensor‡.

---

* p_r is actually a momentum flux density and is different from the ordinary total pressure (stagnation pressure) for steady continuous flow. While p_r remains constant the stagnation pressure changes due to shock transition. For its physical meaning see Fig. 8.1. H_r, however, is the ordinary stagnation enthalpy.

† For the physical meaning of the quantities ϱu and ϱu^2 compare e.g. [4], p. 7.

‡ The general stress tensor

$$\sigma_{ik} = -p\delta_{ik} + \eta\left(\frac{\partial u_i}{\partial x_k} + \frac{\partial u_k}{\partial x_i} - \frac{2}{3}\,\delta_{ik}\,\frac{\partial u_l}{\partial x_l}\right) + \bar{\eta}\delta_{ik}\,\frac{\partial u_l}{\partial x_l} \qquad (i, k, l = 1, 2, 3)$$

yields for $\sigma_{ii} = \sigma_{xx}$ in the case of one-dimensional plane flow $\sigma_{xx} = -p + \hat{\eta}\; du/dx$.

The calculation of the components of the stress tensor in continuum mechanics* shows that even for one-dimensional flow with velocity gradients in the x-direction viscosity has an influence.

An idea of this mechanism may be gained from the continuity equation which requires a change in density for a velocity change, provided that the cross-sectional area of the streamtube remains constant. A one-dimensional compression of a cubic fluid element, however, causes a rotation of the diagonal planes generating a shear of the fluid. Therefore, even in one-dimensional flow both volume changes and a shear deformation of the fluid particle appear in the general case; see Fig. 1.2.

Consequently both types of viscosity, the shear viscosity η and the volume or bulk viscosity $\bar{\eta}$ become effective. Both effects are combined in a single coefficient $\hat{\eta}$,

$$\hat{\eta} = \tfrac{4}{3}\eta + \bar{\eta}.$$

The shear viscosity arises, according to the kinetic theory of gases, from a momentum exchange of fluid particles with different velocities caused by the thermal motion of the molecules. Therefore, the influence of η has to be regarded whenever velocity gradients appear in the flow. The momentum flux caused by the viscosity corresponds to an irreversible transport process.

The bulk viscosity $\bar{\eta}$ is very closely related to the phenomena described by *relaxation* of the inner degrees of freedom. For nonreacting gases the relaxation processes determining the magnitude of $\bar{\eta}$ are primarily due to the rotational degrees of freedom of the molecules (see also section 4). $\bar{\eta} = 0$ holds exactly only for monatomic gases. A fairly general account of the influence of relaxation phenomena in gasdynamics is given in the two volumes of Ref. [7], chapters I, VI, VII, and VIII†. For the influence of $\bar{\eta}$ on the formation of shock waves see also [5], and [13] p. 477.

The *energy theorem* applied to a volume element of the fluid states that the change in the total energy (= internal energy plus kinetic energy) is equal to the work done by the forces acting on the fluid element (= work of the pressure forces and the viscous stresses) plus the energy flux through the surface of the volume element.

---

* See e.g. [4] p. 2; [7] Vol. I p. 69.

† The bulk viscosity, volume viscosity, or dilatational viscosity, described by the coefficient $\bar{\eta}$, is attributed by some authors, e.g. [9], to the rotational inner degrees of freedom only, by others, e.g. [7], [8], [10] to all possible inner degrees of freedom which may be excited including vibration, dissociation, chemical reaction, and ionization. For our purposes it is convenient to regard the coefficient $\bar{\eta}$ as due to the rotational energy only.

The energy balance equation (1.3) derived from the energy theorem therefore expresses that the total specific enthalpy of the fluid flowing through a shock wave remains unchanged. That is, the *energy flux density* has a timewise and spacewise constant magnitude.

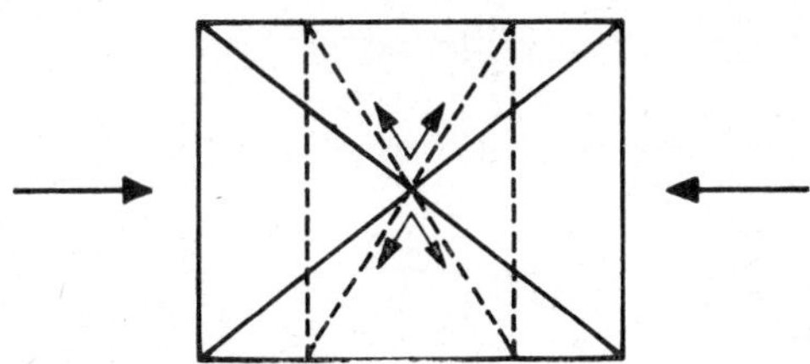

Fig. 1.2 Shear produced in a cubic fluid element due to one-dimensional compression.

This becomes evident if equation (1.3) is written in a slightly different form:

$$\theta E + \theta \frac{u^2}{2} + pu - \hat{\eta} u \frac{du}{dx} - \lambda \frac{dT}{dx} = \text{const.} \tag{1.3'}$$

While the terms in (1.3) are specific energies the terms in (1.3′) are energy flux densities.

The quantities appearing in (1.3′) have the following particular meanings:

E specific internal energy,

$v = \frac{1}{\varrho}$ specific volume,

pv pressure work; in (1.3′) appearing as $\theta pv = pu$,

$\frac{u^2}{2}$ specific kinetic energy,

$\frac{\hat{\eta}}{\varrho} \frac{du}{dx}$ work per unit mass performed by the viscous forces,

$\frac{dT}{dx}$ energy flux due to thermal conduction.

Regarding the constant values p_r and H_r of the flow, we only state here that these values of the flow variables p and H would be obtained if the fluid were brought to rest adiabatically.

1.3 *Transformation of the coordinate system*

The balance equations (1.1) to (1.3) are valid for a steady shock wave, or a coordinate system moving with the shock front.

For a non–steady shock wave propagating with the velocity u_0 in a co-ordinate system fixed with the undisturbed flow in front of the shock wave, the corresponding balance equations are found by means of the proper *Galilei transformation*. This transformation is determined as follows: if

x, t are the coordinates of the moving system, and

x', t' the coordinates of the fixed system,

then

$x' = -x + u_0 t$ (according to Figure 1.1, u_0 points into the negative x' direction),

$t' = t$

hold for the coordinates. For the velocities we obtain

$$\frac{dx'}{dt} = u_0 - \frac{dx}{dt} \quad \text{or} \quad u = u_0 - u'.$$

Using these relations the balance equations take the form:

$$\varrho\,(u_0 - u') = \varrho_0 u_0 = \theta,$$

$$p + \varrho\,(u_0 - u')^2 - \hat{\eta}\,\frac{\partial u'}{\partial x'} = p_0 + \varrho_0 u_0^2 = p_r,$$

$$H + \frac{1}{2}\,(u_0 - u')^2 - \frac{\hat{\eta}}{\varrho}\,\frac{\partial u'}{\partial x'} + \frac{\lambda}{\varrho\,(u_0 - u')}\,\frac{\partial T}{\partial x'} = H_0 + \frac{u_0^2}{2} = H_r.$$

2 Thermodynamic Fundamentals

Regarding the three possible states of aggregation of the matter, the gaseous, liquid, and solid state, the following outlines are intended to describe the gaseous and the liquid state, though the general formulations often apply to the solid state too.

2.1 *Variables of state*

The thermodynamic *state* of the matter considered in the case of thermodynamic equilibrium is assumed to be determined by two independent variables of state or thermodynamic variables. Experience shows that this is true for practically any homogeneous matter in thermodynamic equilibrium. To describe the state of the matter different state variables may be used. These variables may be classified under several aspects. One common classification is the distinction between *intensive* and *extensive* variables of state. For a more detailed discussion see e.g. [5] p. 15, or [13].

Particularly in gasdynamics it becomes appropriate to distinguish between *mechanical* and non-mechanical or *thermal-caloric* thermodynamic quantities. Their properties are discussed in [15] p. 12. Such mechanical quantities are for example:

$$p, \varrho, E, H, a, \Gamma.$$

The meaning of these quantities is explained in the list of symbols.

2.2 *Equations of state*

The thermodynamic *behavior* of matter* is described by certain equations of state which in the case of equilibrium involve three state variables.

There exist *complete* and *incomplete* equations of state. A complete equation of state determines the thermodynamic behavior of the matter comprehensively. With two variables of state prescribed independently all other variables of state may be determined uniquely by the equation of state.

An incomplete equation of state only partly determines the thermodynamic behavior of matter. Not all variables of state can be derived formally by means of the equation.

a) Incomplete equations of state. To the class of incomplete equations of state belong among others the *thermal equations of state*. Those are equations of the form

$$f(p, v, T) = 0. \tag{2.1}$$

With the temperature as a parameter, Eq. (2.1) determines the isothermes (the curves with $T = \text{const}$) in a p–v diagram, that means all the states of a system which may be taken for a prescribed temperature.

The caloric behavior of the matter, however, is not yet known from equation (2.1). To determine the state variables, e.g., "specific internal energy E,

---

* This may be a mixture of several components with a certain gross composition. The components may exist in different phases being in chemical-thermodynamic contact. An explosive and its reaction products e.g. represent the same matter in two different states.

specific enthalpy H, or the specific heat capacity at constant pressure c_p", a further relation is required, namely a *caloric equation of state**, as

$$c_p = c_p(p, T), \tag{2.2}$$

or

$$c_v = c_v(v, T). \tag{2.3}$$

Another incomplete equation of state, which is particularly suitable in gasdynamics, is the *mechanical equation of state*

$$E = E(p, v)$$

or more generally†

$$\mathscr{F}(E, p, v) = 0. \tag{2.4}$$

b) Complete equations of state. Complete equations of state are in particular:

$$E = E(S, v), \tag{2.5}$$

$$H = H(S, p), \tag{2.6}$$

$$F = F(T, v), \tag{2.7}$$

$$G = G(T, p). \tag{2.8}$$

These four functions are called characteristic functions of state. S is the specific entropy, and E the specific internal energy. The specific enthalpy H, the specific free energy F (introduced by HELMHOLTZ), and the specific Gibbs enthalpy G‡ are defined by

$$H = E + pv$$

$$F = E - TS \tag{2.9}$$

$$G = H - TS. \tag{2.10}$$

Further state variables may be defined by means of the total differentials of complete equations of state; see [12] p. 26, 50. For example, according to

---

* In the literature general equations of state like $H = H(p, v)$, or $E = E(v, T)$ giving the specific internal energy or the specific enthalpy as functions of two simple state variables are sometimes also called caloric equations of state. This notation, however, is not uniform and will not be used in this text.

† If the third of the Bethe-Weyl conditions (3.12) is not satisfied, the function $E(p, v)$ is no longer uniquely determined in the whole region. The form (2.4), however, is always applicable.

‡ Sometimes also called free enthalpy and designated by F.

the First and Second Law of thermodynamics and provided that only reversible work and heat transfer occurs, the total differential of (2.5) is

$$dE = T\,dS - p\,dv \tag{2.11}$$

and with

$$d(pv) = p\,dv + v\,dp$$

$$dH = d(E + pv) = T\,dS + v\,dp \tag{2.12}$$

or with

$$d(TS) = T\,dS + S\,dT$$

$$dF = d(E - TS) = -S\,dT - p\,dv. \tag{2.13}$$

Similarly we would find

$$dG = d(H - TS) = v\,dp - S\,dT. \tag{2.14}$$

Each of the characteristic functions (2.5) to (2.8) depends on two of the quantities p, v, T, S as independent variables while the other two appear as partial derivatives. Double differentiation yields all the other important thermodynamic quantities. Expressions like $c_v(v, T)$ or $E(S, v)$ are also sometimes called thermodynamic functions.

The above definitions and relations hold for a non-reacting matter which is in thermodynamic equilibrium.

2.3 *Perfect gases*

a) Thermally perfect gas. One specific form of the thermal equation of state (2.1) is obtained from the empirical laws of BOYLE-MARIOTTE and GAY LUSSAC. This equation of state gives a fairly accurate description of all gases at sufficiently low densities and may be formulated as follows

$$pV = nR_aT = \frac{m}{M_{\mathrm{mol}}} R_aT \tag{2.15}$$

or with

$$\frac{V}{m} = v \quad \text{and} \quad \frac{R_a}{M_{\mathrm{mol}}} = R$$

$$pv = RT. \tag{2.16}$$

The quantities involved are: V volume, n number of moles, R_a universal gas constant, m mass, M_{mol} molar mass, R specific gas constant.

A gas which is described exactly by equation (2.16) is called a *thermally perfect gas* or, briefly, a *perfect gas.*

From equation (2.16) it can be concluded that for a perfect gas the specific internal energy E or the specific heat capacity at constant volume c_v is given by

$$E = E(T),$$

$$c_v = c_v(T).$$

The explicit calculation of the quantities E and c_v, however, requires a further, caloric relation.

In other words, for a thermally perfect gas the following relations hold:

$$\left(\frac{\partial E}{\partial v}\right)_T = 0. \tag{2.17}$$

This gives*

$$c_p - c_v = R, \tag{2.18}$$

$$dE = c_v(T)\, dT, \tag{2.19}$$

$$\Delta E = \int_{T_1}^{T_2} c_v(T)\, dT,$$

$$dH = c_p(T)\, dT, \tag{2.20}$$

$$\Delta H = \int_{T_1}^{T_2} c_p(T)\, dT.$$

b) Calorically perfect gases. If for a thermally perfect gas in addition to equation (2.16)

$$c_p = \text{const}.$$

or on account of (2.18)

$$\frac{c_p}{c_v} = \gamma = \text{const}, \tag{2.21}$$

the gas is called a *calorically perfect gas.* Equations (2.19) and (2.20) then

---

* According to the general definition

$$\left(\frac{\partial E}{\partial T}\right)_v = c_v; \quad \left(\frac{\partial H}{\partial T}\right)_p = c_p.$$

take the simple form

$$E = c_v T; \quad (\Delta E = c_v \Delta T), \tag{2.22}$$

$$H = c_p T; \quad (\Delta H = c_p \Delta T), \tag{2.23}$$

or regarding (2.16) and (2.18)

$$E = \frac{1}{\gamma - 1} pv, \tag{2.24}$$

$$H = \frac{\gamma}{\gamma - 1} pv. \tag{2.25}$$

γ is called *adiabatic exponent* on accourt of Eq. (2.39).
Equation (2.24) or (2.25) may be called accordingly the mechanical equation of state for a calorically perfect gas.

The γ values of calorically perfect gases are found to be approximately equal to the ratio of integer numbers with values according to the number of atoms in the molecule. The explanation is given by the kinetic theory of gases and the statistical mechanics. The specific heat depends on the number of excited degrees of freedom* of the molecules and on a their energy. The classical theory distinguishes between:

a) Translational degrees of freedom.
b) Rotational degrees of freedom.
c) Vibrational degrees of freedom.

Each degree of freedom contributes a kinetic energy equal to $R/2$ to the specific heat c_v, and the vibrational degrees of freedom an additional potential energy $R/2$ for each mode of vibration. The quantity $f = 2c_v/R$ is therefore in general not equal to the number of the mechanical degrees of freedom. We shall call f the number of thermodynamic degrees or quasi degrees of freedom. f increases in general with the number of atoms of the molecules. For

monatomic gases	$f = 3$
diatomic gases	$f = 3 + 2 + 2$
triatomic gases,	
linear configuration	$f = 3 + 2 + 8$
angular configuration	$f = 3 + 3 + 6.$

* For further discussion compare for example [11] p. 43.

The translational and rotational degrees of freedom are already completely excited at fairly low temperatures (for air e.g. above 50 K). The vibrational degrees of freedom on the other hand become excited only at higher temperatures. We assume, however, at first $f = \text{const}$. For thermal perfect gases according to (2.18) the following relations hold:

$$c_v = \frac{f}{2} R, \quad c_p = \left(\frac{f}{2} + 1\right) R,$$

$$\frac{c_p}{c_v} = \gamma = 1 + \frac{2}{f}, \tag{2.21'}$$

$$f = \frac{2c_v}{R} = \frac{2C_v}{R_a}, \quad f + 2 = \frac{2c_p}{R} = \frac{2C_p}{R_a},$$

(C_p and C_v are molar heat capacities).
Because $R_a = 1.987$ cal degree^{-1}mole^{-1} the numerical value of f is approximately equal to C_v, provided that C_v is measured in cal degree^{-1} mole^{-1} (see also Table 4.1).

Equations (2.24), (2.25), when written in terms of f, read

$$E = \frac{f}{2} pv, \tag{2.24'}$$

$$H = \left(1 + \frac{f}{2}\right) pv. \tag{2.25'}$$

By means of equation (2.24′) or (2.25′) a number f may also be defined for imperfect gases. f, the number of "quasi degrees of freedom", is then, however, no longer constant and (2.21′) is no longer valid.

A few further thermodynamic quantities and relations which will be used in the text are:

The *velocity of sound* (see Sec. 3.5) as defined by

$$a^2 = \left(\frac{\partial p}{\partial \varrho}\right)_S. \tag{2.26}$$

A *generalized adiabatic exponent* Γ ([15], p. 14)

$$\Gamma = \frac{a^2}{pv} = \left(\frac{\partial \ln p}{\partial \ln \varrho}\right)_S = \left(\frac{\partial H}{\partial E}\right)_S = \frac{v}{p} \frac{(\partial E/\partial v)_p + p}{(\partial E/\partial p)_v}. \tag{2.27}$$

Γ as defined above is a *local* adiabatic exponent and in general

$$\Gamma \neq \frac{c_p}{c_v}; \quad \Gamma = \Gamma(p, v). \tag{2.28}$$

However, for a perfect gas always $\Gamma = c_p/c_v$.
In the vicinity of a particular state point (p_0, v_0) along the isentropic curves we have approximately

$$pv^{\Gamma(p_0, v_0)} \approx \text{const.}$$

The *thermal expansion coefficient* α which is defined by

$$\alpha = \frac{1}{v}\left(\frac{\partial v}{\partial T}\right)_p = \left(\frac{\partial \ln v}{\partial T}\right)_p. \tag{2.29}$$

2.4 *Thermodynamic classification of the flow processes*

Regarding the different states assumed in a fluid flow [19] we will distinguish between:

a) processes,
b) changes of state,
c) space-time variations corresponding to a certain field distribution.

Furthermore, we will distinguish according to the above classification between

α) a system which undergoes a process,
β) a homogeneous fluid particle* which is subjected to a series of changes of state,
γ) a flow field with a certain space-time distribution of material-kinematic states.

The system considered can be composed of different parts and may comprise a total flow process with in general non-uniform states (compare e.g. Fig. 1.1) whereas the fluid particle is assumed to be sufficiently small so that its state can be considered as spatially homogeneous, i.e. dependent only on time.

The general changes appearing in the flow fields classified above are furthermore subject to certain thermodynamic conditions.

The relations between the different terms are given in Table 2.1.

---

* An "air particle" of e.g. 10^{-3} mm^3 would still contain in the order of 10^{13} molecules.

Table 2.1

system, process	reversible	irreversible
	adiabatic	diabatic
fluid particle change of state	isentropic	non-isentropic
flow-field distribution	isentropic	non-isentropic
	homentropic	non-homentropic

For isentropic flow $DS/Dt = 0$* while for homentropic flow $S =$ constant.

A classification of the types of flow can be obtained by selecting particular thermodynamic conditions. This classification may lead to certain models for a simplified description of the real flow process.

For one-dimensional plane steady flow the following processes are important.

a) Adiabatic processes A thermodynamic process which takes place in a thermally isolated system is called adiabatic. There is no possible heat transfer through the boundaries of the system. Work performed by or on the system as well as internal development of heat and an entropy change are still possible.

We may therefore distinguish between two types of adiabatic processes,

a) *adiabatic irreversible processes*—the balance shows an entropy increase according to nonisentropic changes of state within the system, and
b) *adiabatic reversible processes*—the total entropy of the system remains constant due to isentropic changes of state which may occur only.

b) Non-adiabatic (diabatic) processes. Those processes may involve a heat transfer through the boundaries of the system.

This classification may be explained by means of some examples.

Chemical reactions may occur as adiabatic processes. Heat may be released by energy transformation with the heat, however, being kept within the system. Reactions with a shifting state of equilibrium can be reversible (in the limiting case) as a series of isentropic changes of state. Reactions with non-equilibrium states like combustions and detonations, however, are irreversible processes involving nonisentropic changes of state.

---

* $D/Dt = \partial/\partial t + u\,(\partial/\partial x)$.

Sound waves and shock waves may occur both as adiabatic processes according to the above definition. Whereas sound waves, however, produce isentropic changes of state (at least within the approximation of linear acoustics in an ideal fluid flow)*, the flow through a shock wave is basically an irreversible process. Shock transition causes an entropy increase.

Non-adiabatic processes, particularly isothermal processes are used to define reaction energies.

The most important thermodynamic processes which are considered in the following are summarized in Table 2.2.

Table 2.2

	Condition	Process	Example
A	$\delta Q_e = 0$ $\delta Q_i = 0;\quad dS = 0$	adiabatic-reversible	sound wave
B	$\delta Q_e = 0$ $\delta Q_i = 0;\quad dS > 0$ $\hat{\eta} \neq 0$	adiabatic-irreversible without heat release, viscous	shock wave
C	$\delta Q_e = 0$ $\delta Q_i < 0;\quad dS > 0$ $\hat{\eta} = 0$	adiabatic-irreversible with heat release, non-viscous	combustion

Since we are mainly interested in reactive flow, we will distinguish between an increase of heat,

a) due to a heat addition from external sources described by δQ_e, and
b) due to heat production caused by chemical reactions described by $-\delta Q_i$.

For adiabatic processes we consequently have $\delta Q_e = 0$. The combination of processes B and C corresponds to a detonation: $B + C =$ Detonation (ZND model, see Chapter III). For steady flow with constant cross-sectional area of the stream tubes the process C corresponds to a Rayleigh process (for definition see Section 8).

* An ideal fluid (gas or liquid) is one which is neither heat conducting nor viscous. To distinguish an ideal fluid from a perfect one defined above and because viscosity has an effect only if the fluid is in motion it seems more appropriate to call it an ideal flow.

A fluid particle in an ideal fluid (flow) can obviously be subject only to adiabatic changes of state. If only equilibrium states occur the change is isentropic. Therefore sometimes, adiabatic flow is set identical to isentropic flow.

For the description of the different processes the First and Second Laws of thermodynamics are used. For the closed system the following relation holds

$$dE = \delta Q_a + \delta A. \tag{2.30}$$

δQ_e is the heat added to the system from external sources. δA is the total external work performed on the system, very often represented by $\delta A = -p\,dv$. For the changes of state of a fluid particle undergoing a process within the system we have

$$T\,dS = dE + p\,dv = dH - v\,dp. \tag{2.31}$$

Adiabatic irreversible processes (*B* Processes)

$$\delta Q_e = \delta Q_i = 0; \quad \hat{\eta} \neq 0; \quad \Delta S > 0.$$

Shock waves, for example, belong to this type of processes. The increase of entropy in the system is determined by the magnitude of the dissipation; see e.g. [5], page 66. The change of state is described by

$$dS = dS_{\text{rev}} + dS_{\text{irr}} = \frac{dE + p\,dv}{T} + dS_{\text{irr}} \tag{2.32}$$

where dS_{rev} is the reversible change in entropy which is zero for a whole cycle, and dS_{irr} is the irreversible increase of entropy due to entropy production.

If $\delta Q_i \neq 0$, heat addition or substraction due to chemical reaction has to be regarded; see Section 5.

Adiabatic reversible processes (*A* processes)

$$\delta Q_e = \delta Q_i = 0; \quad \hat{\eta} = 0; \quad \Delta S = 0. \tag{2.33}$$

The changes of state are determined by

$$dS = \frac{dE + p\,dv}{T} \tag{2.34}$$

which is zero on integration over a whole cycle. For a thermally perfect gas we may write according to (2.19) and (2.16)

$$dS = c_v \frac{dT}{T} + R\frac{dv}{v} = c_v\,d\ln T + R\,d\ln v; \tag{2.35}$$

which, by integration in the case of a calorically perfect gas gives according to (2.21):

$$S - S_0 = \Delta S = c_v \ln \frac{T}{T_0} + R \ln \frac{v}{v_0} = c_v \ln \left[\left(\frac{T}{T_0} \right) \left(\frac{v}{v_0} \right)^{\gamma - 1} \right], \qquad (2.36)$$

or with (2.16)

$$\Delta S = c_v \ln \left(\frac{p}{p_0} \right) \left(\frac{v}{v_0} \right)^{\gamma}. \qquad (2.37)$$

Therefore for the changes of state of a calorically perfect gas enclosed in an adiabatic system we may write

$$\frac{p}{p_0} = \left(\frac{v}{v_0} \right)^{-\gamma} e^{\Delta S / c_v} = \left(\frac{\varrho}{\varrho_0} \right)^{\gamma} e^{\Delta S / c_v} \qquad (2.38)$$

and if ΔS is identical to zero

$$p v^{\gamma} = p_0 v_0^{\gamma} = \text{const.} \qquad (2.39)$$

Equation (2.39) is called the *Poisson equation* and its representation in a p–v diagram, the *isentrope*.

Isothermal diabatic processes

$$\delta Q_e \neq 0; \quad \delta Q_i \neq 0 \qquad (2.40)$$

e.g. determination of reaction heats: $\delta Q_e = -\delta Q_i < 0$.

3 The Shock Adiabat

The system of balance equations (1.1) to (1.3) has besides the identical solution a solution $u_2 \neq u_0$ for the flow behind the shock. The shock layer becomes a real part of the flow and the question of its structure arises.

The shock wave can be described from two different points of view, depending on whether the transition zone or the flow before and behind this zone is of dominant importance.

Viewpoint a) The shock layer which has in general an extension of the order of a few mean free paths of the molecules is considered as a discontinuity surface in the mathematical sense. The description is restricted to the flow regions before and behind the shock layer where all gradients become vanishingly small; see Fig. 3.1.

Viewpoint b) The shock transition layer itself is the region of interest and the subject of the investigation. The flow outside of the transition layer is assumed to be approached asymptotically for $x \to \pm\infty$; see Fig. 3.2.

The shock wave has been investigated based on viewpoint a) by RANKINE and more generally by HUGONIOT. Most of the historical references are given in [1].

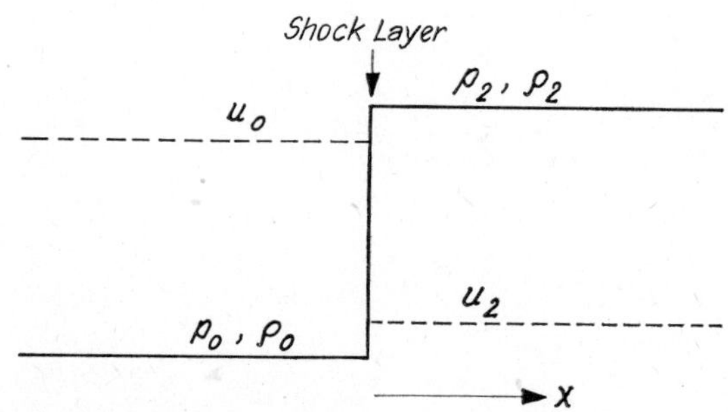

Fig. 3.1 The shock front as a discontinuity.

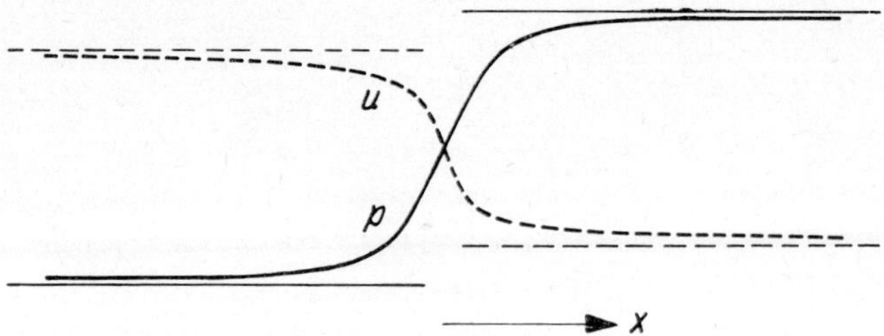

Fig. 3.2 The shock front as a transition region.

The fundamental discussion of shock waves and detonation waves made by R. BECKER in 1922, Ref. [16], was based on viewpoint b). The aim of these investigations was to obtain an idea about the structure of the shock transition layer.

BECKER's work rests upon the premises made in continuum mechanics. For regions of so small extent as in the case of a shock transition zone with finite changes of state the premises of continuum mechanics, however, are no longer guaranteed. Heat conduction and viscosity will cause non-equilibrium states and the duration of the processes leading to equilibrium states becomes increasingly important. The proper approach to investigate the structure of a shock front would be by means of the kinetic theory of gases.

3.1 *The shock relations*

Based on viewpoint a) outlined above, a number of relations describing the flow on both sides of the shock can be derived from the balance equations, disregarding the particular flow inside the shock transition zone. For this description of the flow through a shock wave to be valid it is only required that gas kinetic equilibrium* be established in the flow regions 0 and 2, see Fig. 1.1. Resulting from this condition the velocity and temperature gradients vanish in the flow regions 0 and 2, and the balance equations simplify to

$$\varrho_2 u_2 = \varrho_0 u_0 = \theta \tag{3.1}$$

$$p_2 - p_0 + \varrho_2 u_2^2 - \varrho_0 u_0^2 = p_2 - p_0 + \theta\,(u_2 - u_0) = 0, \tag{3.2}$$

$$H_2 - H_0 + \tfrac{1}{2}\,(u_2^2 - u_0^2) = 0. \tag{3.3}$$

With the following notations

$$\Delta u = u_2 - u_0, \quad \Delta p = p_2 - p_0,$$

$$\bar{u} = \frac{u_2 + u_0}{2}; \quad \bar{v} = \frac{v_2 + v_0}{2}, \tag{3.4}$$

equations (3.1) to (3.3) can be brought into the simple form

$$\Delta u - \theta\,\Delta v = 0; \quad \bar{u} = \theta\bar{v}, \tag{3.5}$$

$$\Delta p + \theta\,\Delta u = 0, \tag{3.6}$$

$$\Delta H + \bar{u}\,\Delta u = 0. \tag{3.7}$$

The balance equations in form of equations (3.1) to (3.3) or (3.5) to (3.7) are usually called the *shock relations.*

3.2 *The Rayleigh line*

By eliminating the velocity change Δu from equations (3.5) and (3.6) a linear v–p relation can be derived for the shock transition:

$$\Delta p + \theta^2\,\Delta v = 0. \tag{3.8}$$

This relation describes a straight line $\mathscr{R}$† in a p–v diagram; see Fig. 3.3.

---

* The thermodynamic-chemical equilibrium can be for instance a metastable equilibrium with frozen reaction; compare also Section 4.

† In Ref. [28], p. 74, called Michelson line.

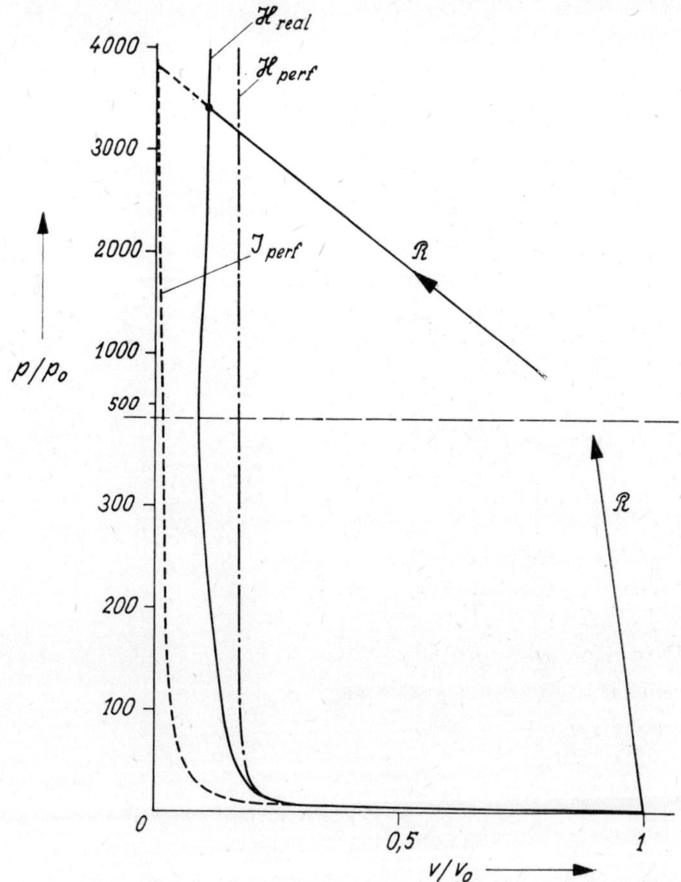

Fig. 3.3 Hugoniot curve for real air according to Ref. [17] (two different ordinate scales).

Equation (3.8) which is called the mechanical condition for shock transition (see [1] p. 130) requires no specification of the energy balance. Therefore, this relation generally holds for processes which satisfy equations (3.1) and (3.2). See also Section 8.

3.3 *The Hugoniot equation*

If we eliminate the flow velocities u_2 and u_0, that means the quantities $\bar{u}$ and Δu, from the energy balance equation (3.7) by means of equations (3.5) and (3.6) we obtain the following relation

$$\Delta H = \bar{v}\, \Delta p, \tag{3.9}$$

or with E instead of H, compare equations (2.11) and (2.12),

$$\Delta E = -\bar{p}\,\Delta v. \tag{3.10}$$

Equation (3.9) or (3.10) was introduced by H. HUGONIOT and is usually called the *Hugoniot equation.* This equation which has the two equivalent forms (3.9) and (3.10) relates the thermodynamic states on both sides of a compression or dilatation wave of finite amplitude outside the transition zone and does not depend on the velocities involved. According to its derivation the Hugoniot equation is valid under the fairly general conditions that the influence of viscosity and heat conduction can be neglected outside the (generally small) transition zone. Flows with chemical reactions, e.g., are not excluded. For those types of flow the equations of state describing the initial and final states of the fluid are determined by different functions and the explicit form of the Hugoniot equation is then more appropriate. The two equivalent forms of the explicit Hugoniot equation are

$$\begin{aligned} H_2 - H_0 &= \tfrac{1}{2}(v_2 + v_0)(p_2 - p_0), \\ E_2 - E_0 &= -\tfrac{1}{2}(p_2 + p_0)(v_2 - v_0). \end{aligned} \tag{3.11}$$

A graph of the Hugoniot relation in a p–v diagram for a particular fluid is called *Hugoniot curve* or *shock adiabat.*

The shock adiabat $\mathscr{H}$ determines all the possible final states 2 of a gas which can be obtained from a given initial state 0 due to shock transition. Which particular final state is taken in each case is determined by a further quantity, e.g., the initial flow velocity u_0. If no particular final states are referred to, we write the coordinates v, p instead of v_2, p_2, see Figs. 3.3 to 3.7. The initial point on the shock adiabat is determined by the thermodynamic initial state 0.

The initial state and the flow velocity in front of the shock wave u_0 determine the Rayleigh line $\mathscr{R}$. The points of intersection between the Rayleigh line and the shock adiabat determine the initial and final states of the gas.

This behavior expresses only the aforementioned fact that the balance equations (3.5) to (3.7) admit a further solution besides the identical solution. The points of intersection between the Hugoniot curve and the Rayleigh line correspond to these two solutions of this system of equations.

The process which actually occurs between (0) and (2) when the gas is passing through the shock front cannot be described by the Hugoniot equation or the mechanical shock conditions. This is according to standpoint a) on which our discussion is based. The shock transition zone where viscosity and heat conduction are effective is excluded from the above discussion.

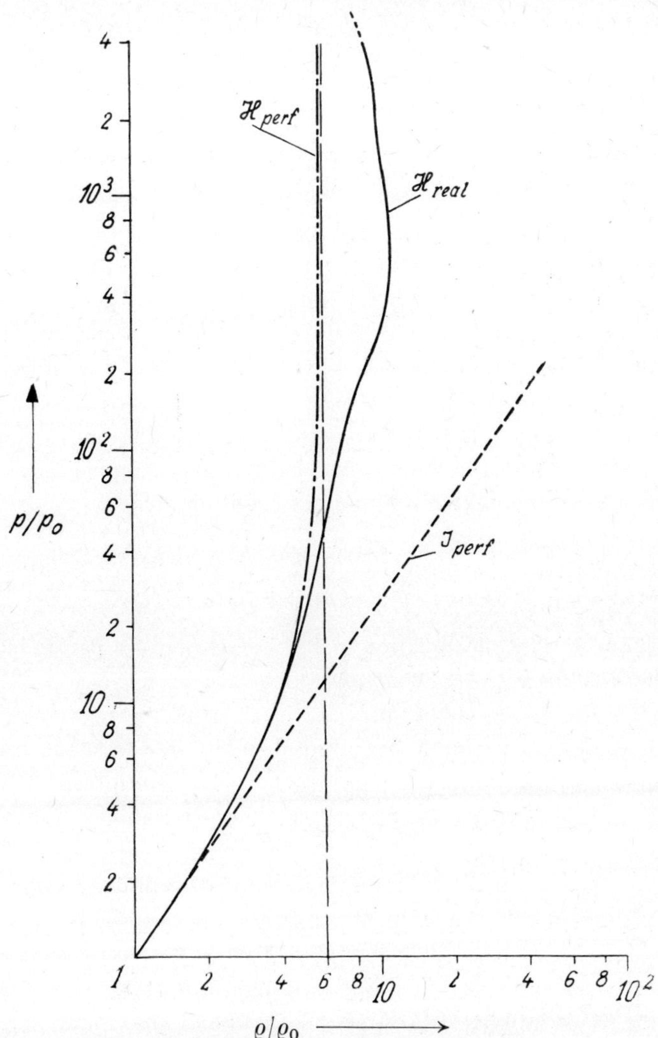

Fig. 3.4 Hugoniot curve and isentrope for air (redrawn from Fig. 3.3).

The shape of the Hugoniot curve is determined by the thermodynamic characteristics of matter. For substantial changes of state, as in strong shock waves for example, the deviations from the characteristics of a perfect gas become more and more pronounced. This becomes evident in Figs. 3.3 and 3.4 where a real Hugoniot curve $\mathscr{H}_{\text{real}}$ with data taken from [17] is being compared with the corresponding curve for a calorically perfect gas with the same initial values at (0).

Shock waves in fluids (liquids and gases) have been investigated theoretically under very general conditions by H. A. BETHE (1942) and H. WEYL (1949). These authors have established sufficient thermodynamic conditions for a "normal behavior" of the fluid. These conditions are usually combined in the following form (compare [1] and [3] p. 427):

$$\begin{aligned} &\text{a)} \quad \left(\frac{\partial p}{\partial v}\right)_S < 0, \\ &\text{b)} \quad \left(\frac{\partial^2 p}{\partial v^2}\right)_S > 0, \\ &\text{c)} \quad \left(\frac{\partial p}{\partial S}\right)_v > 0. \end{aligned} \tag{3.12}$$

Condition a) is satisfied for all known materials and implies the existence of a real velocity of sound according to (2.26). Condition b) is, according to Ref. [16], substantial for the stability of shock waves and excludes the existence of rarefaction shocks. This condition probably holds for all fluids*. Condition c) is less essential and might occasionally not be satisfied near phase limits (anomaly of water below 4 °C).

3.4 *Change of entropy along the shock adiabat (area rule)*

A comparison of the shock adiabat (or Hugoniot curve) with the isentrope curve $\mathscr{I}_0$ through the initial point (0) shows that the shock adiabat intersects the isentrope at (0), the first and second derivatives being identical. The first difference occurs in the third derivative; the intersection of the two curves is of third order (Fig. 3.5).

The order of magnitude of the entropy changes along the shock adiabat can be derived as follows: starting with relation (2.11)†

$$T\,dS = dE + p\,dv$$

and the Hugoniot equation (3.10)

$$\Delta E = -\bar{p}\,\Delta v,$$

$$dE = -\tfrac{1}{2}\,\Delta v\,dp - \bar{p}\,dv$$

---

* For solids this condition is often violated.

† Relation (2.11) is valid under the same conditions as the Hugoniot equation.

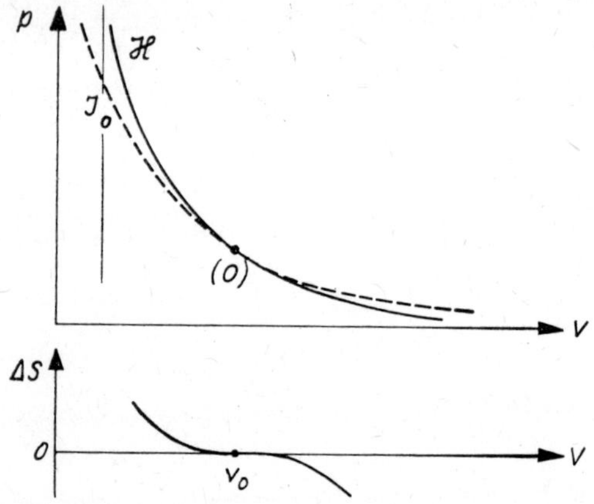

Fig. 3.5 Shape of the isentrope and shock adiabat and the entropy change at (0), from Ref. [16].

we obtain

$$T\,dS = \tfrac{1}{2}\,\Delta p\,dv - \tfrac{1}{2}\,\Delta v\,dp = dF; \tag{3.13}$$

$dF = T\,dS$ may be interpreted as the differential of an area as drawn in Fig. 3.6. The entropy S changes along $\mathscr{H}$ as the area F passed over by the Rayleigh line, (area rule):

$$\begin{aligned} dF &= -\tfrac{1}{2}(\Delta v + dv)(\Delta p + dp) + \tfrac{1}{2}\Delta v\,\Delta p + \Delta p\,dv + \tfrac{1}{2}\,dp\,dv \\ &= -\tfrac{1}{2}\,\Delta v\,dp + \tfrac{1}{2}\,\Delta p\,dv. \end{aligned}$$

A right hand sweep of the Rayleigh line corresponds to an entropy increase and vice versa.

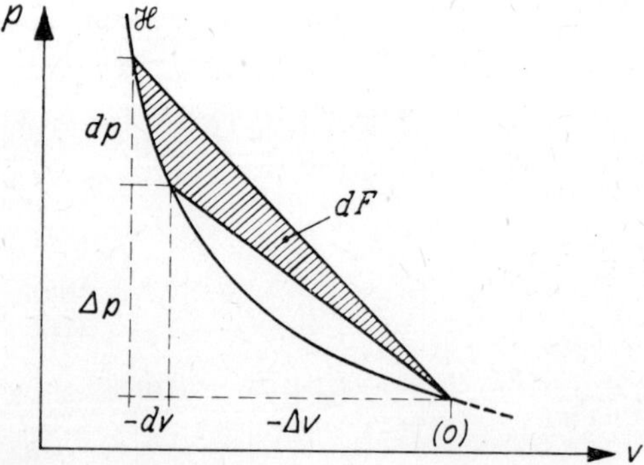

Fig. 3.6 The area rule.

If we differentiate (3.13) twice with respect to v we obtain

$$\frac{dT}{dv}\frac{dS}{dv} + T\frac{d^2S}{dv^2} = -\frac{1}{2}\Delta v\frac{d^2p}{dv^2}, \tag{3.14}$$

$$\frac{d^2T}{dv^2}\frac{dS}{dv} + 2\frac{dT}{dv}\frac{d^2S}{dv^2} + T\frac{d^3S}{dv^3} = -\frac{1}{2}\Delta v\frac{d^3p}{dv^3} - \frac{1}{2}\frac{d^2p}{dv^2}. \tag{3.15}$$

In the vicinity of (0) Δp and Δv approach 0 and therefore

$$\left(\frac{dS}{dv}\right)_0 = \left(\frac{d^2S}{dv^2}\right)_0 = 0, \quad \left(\frac{d^3S}{dv^3}\right)_0 = -\frac{1}{2T_0}\left(\frac{d^2p}{dv^2}\right)_0,$$
$$dS = -\frac{1}{12T_0}\left(\frac{\partial^2p}{\partial v^2}\right)_s dv^3. \tag{3.16}$$

The entropy change is of third order with respect to dp or dv. The sign of the change is determined by (3.12b), see Fig. 3.5.

3.5 *Sound waves*

The square of the propagation velocity of a small pressure disturbance in a compressible fluid is*

$$a^2 = \left(\frac{\partial p}{\partial \varrho}\right)_s.$$

That means the change of state due to such a pressure wave is along an isentrope curve. Pressure waves which cause an isentropic change of state belong to the field of linear, absorption-free acoustics. This is approached if the deviations from the equilibrium state are small and if the influences of heat conduction and viscosity may be neglected. If absorption has to be taken into account, the changes of state are no longer perfectly isentropic and in principle the sound velocity is no longer exactly given by (2.26); compare Section 4. Practically, however, if classical absorption (heat conduction and shear viscosity) is the only one effective, the changes are so small that the equilibrium conditions are applicable.

From the above discussion of the entropy change along the shock adiabat, it follows that weak shock waves propagate with the velocity of sound and, as the intensity decreases, the characteristics of a sound wave are more and more approached.

---

* Cf. (2.26); for derivation see e.g. [3] p. 354; [8] p. 245.

From the relation between shock adiabat and Rayleigh line we may immediately derive a necessary condition for the existence of a shock wave. To obtain two non-identical points of intersection it is required that

$$\theta^2 = \left|\left(\frac{\Delta p}{\Delta v}\right)_{\mathscr{R}}\right| > \left|\left(\frac{dp}{dv}\right)_{\mathscr{H}}\right|_0 . \tag{3.17}$$

This means that the flow velocity u_0 in front of the shock wave must be greater than the velocity of sound a_0:

$$u_0 > a_0 .$$

If the incident flow velocity becomes sonic, that means $u_0 = a_0$, the Rayleigh line is a tangent to the Hugoniot curve at the initial point (0). At this tangential point the slopes of the isentrope, the Hugoniot curve, and the Rayleigh line are equal and we may write there

$$\left(\frac{\partial p}{\partial v}\right)_s = \left(\frac{\partial p}{\partial v}\right)_{\mathscr{H}} = \left(\frac{\partial p}{\partial v}\right)_{\mathscr{R}} = -\theta^2. \tag{3.18}$$

From (2.26) together with the condition for isentropic changes of state in a perfect gas according to (2.39) we obtain for the velocity of sound a

$$a^2 = \gamma \frac{p}{\varrho} = \gamma p v \tag{3.19}$$

where $\gamma = \gamma(T)$ is in general a function of the temperature T. Regarding (2.16) Eq. (3.19) writes

$$a^2 = \gamma R T. \tag{3.20}$$

For a perfect gas the velocity of sound is only a function of the temperature.

For a non-perfect gas relation (3.19) can be written in the same form as

$$a^2 = \Gamma p v, \tag{3.21}$$

with Γ being a generalized local adiabat exponent; compare the definition (2.27).

Only for a perfect gas

$$\Gamma \equiv \gamma .$$

3.6 *Shock waves in a colorically perfect gas*

For a calorically perfect gas the shock adiabat is described by the relation

$$\frac{v}{v_0} = \frac{(p/p_0) + (\gamma + 1)/(\gamma - 1)}{(\gamma + 1)/(\gamma - 1)\,(p/p_0) + 1} \tag{3.22}$$

regarding Eqs. (2.24) and (2.25). The corresponding relations for p and T are:

$$\frac{p}{p_0} = \frac{(\gamma + 1)/(\gamma - 1) - (v/v_0)}{(\gamma + 1)/(\gamma - 1)(v/v_0) - 1}; \tag{3.23}$$

$$\frac{T}{T_0} = \frac{(p/p_0)[1 + (\gamma - 1)/(\gamma + 1)(p/p_0)]}{(p/p_0) + (\gamma - 1)/(\gamma + 1)}. \tag{3.24}$$

(3.22) or (3.23) is the equation of a hyperbola with the asymptotes

$$\begin{aligned} \frac{v}{v_0} &= \frac{\gamma - 1}{\gamma + 1} = \frac{1}{f + 1} \qquad (p \to \infty), \\ \frac{p}{p_0} &= -\frac{\gamma - 1}{\gamma + 1} = -\frac{1}{f + 1} \qquad (v \to \infty). \end{aligned} \tag{3.25}$$

Regarding (2.21) which gives the ratio of the specific heats in terms of the molecular quasi degrees of freedom, Eq. (3.22) becomes

$$\frac{v}{v_0} = \frac{(p/p_0) + f + 1}{(f + 1)(p/p_0) + 1}. \tag{3.26}$$

For example, if $\gamma = 1.4$, then $f = 5$ and Eq. (3.26) may be written

$$\frac{v}{v_0} = \frac{(p/p_0) + 6}{6(p/p_0) + 1} \tag{3.27}$$

with the particular asymptotes: $v/v_0 = \frac{1}{6}$, $p/p_0 = -\frac{1}{6}$. These values are a good approximation for air under normal conditions. The shape of the Hugoniot curve together with the isentrope in the case $\gamma = 1.4$ is shown in Fig. 3.7.

Eqs. (2.24), (2.25) and (3.8) to (3.10) yield the following additional relations

$$\theta^2 = -\frac{\Delta p}{\Delta v} = \gamma \frac{\bar{p}}{\bar{v}}, \quad \frac{\Delta p}{\bar{p}} = -\gamma \frac{\Delta v}{\bar{v}} = \gamma \frac{\Delta \varrho}{\bar{\varrho}}. \tag{3.28}$$

In formulas (3.28) [Eichelberg, v. Karman] as well as in (3.8) to (3.10) in the limit of weak shocks ($\Delta \to d$) the transformation to the acoustic isentropic formulas becomes evident.

If the slope of the Rayleigh line through (0) becomes smaller than $\varrho_0^2 a_0^2$, the final state is determined by a point of intersection on the lower branch of the shock adiabat below the initial point (0). In this case the flow velocity

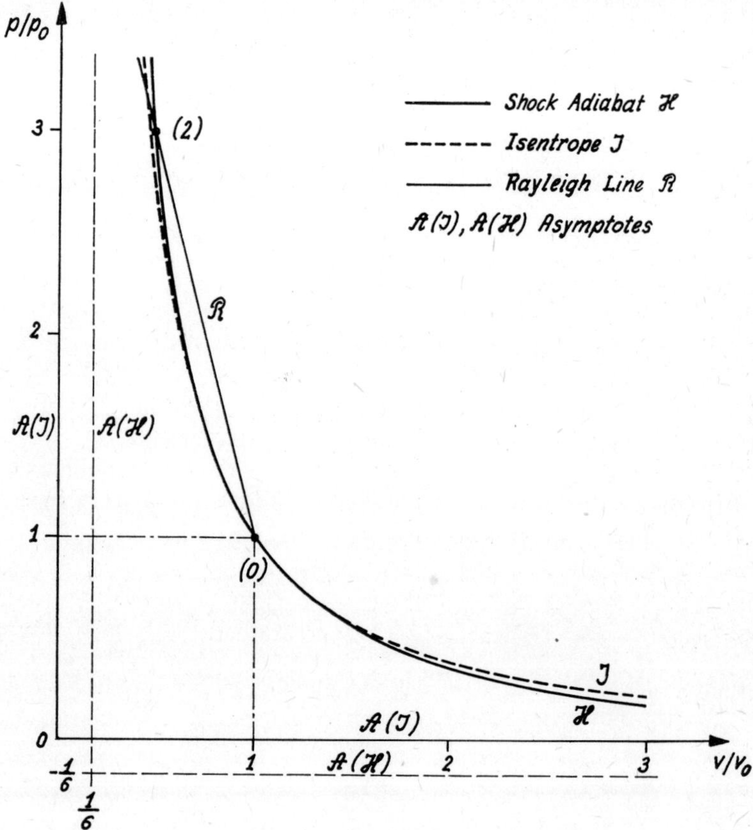

Fig. 3.7 Adiabats for calorically perfect gas ($\gamma = 1.4$).

in the undisturbed region is subsonic, $u_0 < a_0$, and the transition would correspondingly represent a "rarefaction discontinuity". It can be shown by means of the second law of thermodynamics that such a "rarefaction shock" cannot be stable; see e.g. [18].

CHAPTER II

Shock Waves with Relaxation

4 Nonequilibrium Flow

THE INSTANTANEOUS establishment of chemical-thermodynamic equilibrium immediately behind the shock front is in general not achieved. So far we have made the assumption that, in the regions (0) and (2) of Fig. 1.1, equilibrium conditions are given. In the shock layer represented by region (1), nonequilibrium conditions have to be assumed, of course, in any case. However, even in the region behind the shock front, we may expect certain reaction processes caused by the changes of state due to shock transition with reaction rates slow compared with the shock transition [20] [21] [22].

Nonequilibrium states and the effects of finite reaction times are particularly important for flow phenomena with heat production like combustion processes.

In the formal description of flow phenomena with certain reactions the reaction process may be understood as a very general transition of energy not necessarily restricted to chemical reactions only*.

4.1 *Relaxation*

The term *relaxation* will be used here in the general sense describing processes leading to thermodynamic equilibrium of a substance.

To describe the state of a matter taking relaxation phenomena into account we need, in addition to the two thermodynamic quantities, at least one further independent variable ξ determining the progress of the reaction. ξ is called the *extent of reaction* or *reaction variable.* The substance being in a nonequilibrium state is now described by at least three independent variables of state, e.g. p, v, and ξ. The equation of state (2.4) describing the mechanical-thermodynamic behavior of the matter has now the form

$$E = E(p, v, \xi)$$

* References containing a more extensive treatment of the subject matter are e.g., [5], [7] or [9].

or

$$H = H(p, v, \xi). \tag{4.1}$$

Approaching the equilibrium state, ξ tends towards a value $\tilde{\xi}$ which depends on p and v. Therefore, in equilibrium, the state of the system is again determined by only two variables e.g., p and v. We write

$$\xi \rightarrow \tilde{\xi}(p, v).$$

If the changes of state occur sufficiently slowly compared with reaction time, the reaction variable follows the changes almost completely, that means we always find $\xi \simeq \tilde{\xi}(p, v)$. In those cases we may set

$$\xi \equiv \tilde{\xi}(p, v) \quad \text{“shifting equilibrium”}.$$

These conditions are found for example in sound waves at sufficiently low frequencies*. The common methods of equilibrium thermodynamics (see Ref. [19], p. 12–13) are then applicable and the equation of state has the form (see below (4.15))

$$H = H^e(p, v) = H(p, v, \tilde{\xi}(p, v)).$$

In the general case of nonequilibrium processes several independent variables $\xi_1, \xi_2, \ldots, \xi_n$ are required to describe the state of the fluid. The proper equation of state $H(p, v, \xi_1, \xi_2, \ldots, \xi_n)$ may be written briefly as

$$H = H(p, v, \xi_i). \tag{4.2}$$

Regarding the description of the reaction process we will distinguish between four different levels of information:

1) A description by means of mechanical-thermodynamic quantities like p, ϱ, E, H, ξ_i. These quantities are sufficient to define the instantaneous state of the system from a gasdynamic point of view.

2) A description by means of thermal-caloric-thermodynamic quantities T, S, μ (chemical potentials), K (equilibrium constants). This description yields information about affinity and position of the equilibrium.

3) A phenomenological-reaction-kinetic description by means of reaction rates given by

$$\frac{d\xi_j}{dt} = r_j(p, v, \xi_i).$$

* Depending on the relaxation time of the non-equilibrium processes, these frequencies can be extremely high.

4) A description by elementary-reaction kinetic relations involving activation energies E_{act} and reaction-rate constants for forward and back elementary reactions, k_f and k_b. By means of these quantities detailed information is obtained about the progress of the reaction.

The higher levels of information as listed in the above order also require in general, a more detailed knowledge of the quantities determining the reaction process.

The reaction rate

$$\frac{d\xi}{dt} = r(p, v, \xi)$$

or in the general case

$$\frac{d\xi_j}{dt} = r_j(p, v, \xi_i) \tag{4.3}$$

is a function of the variables of state. This function is very often not known exactly and a model has to be constructed which permits the most accurate description.

The equilibrium states $\tilde{\xi}_i$ are characterized by the vanishing of all reaction rates

$$r_j(p, v, \tilde{\xi}_i(p, v)) = 0.$$

For a simplified description we will in general use the information levels 1) and 3).

4.2 *Forms of energy of a gas*

The internal energy of a gas is composed in the most general case by the following parts:

1) Translational energy (tr)
2) Rotational energy (rot)
3) Vibrational energy (vib or osc)
4) Chemical reaction energy (chem)
5) Electronic excitation energy (exc)
6) Ionization energy (ion).

The sequence 1 to 6 of the energy forms characterizes approximately the direction of increasing slowness of equilibrium approach.

Processes which proceed within the shock front may be considered as infinitely rapid and the shock front approximated as a discontinuity in the mathematical sense. The equilibrium adjustment of the translational energy and the rotational energy belong to these processes due to their extremely short relaxation time. The classical transport phenomena (diffusion, heat conduction, and viscosity) are governed by these rapid processes.

Processes which proceed slowly compared with the transition through the shock front are in general those listed as parts 3 to 6 of the internal energy. In Ref. [27] Hugoniot curves are described for diatomic gases, particularly nitrogen, which would be found for a stepwise adjustment of the equilibrium according to the above listed order of energy forms.

4.3 *The shock-induced relaxation layer*

For the shock transition of a gas regarding the relaxation of a reaction equilibrium we assume for the further discussion the following subdivision of the flow region, see Fig. 4.1.

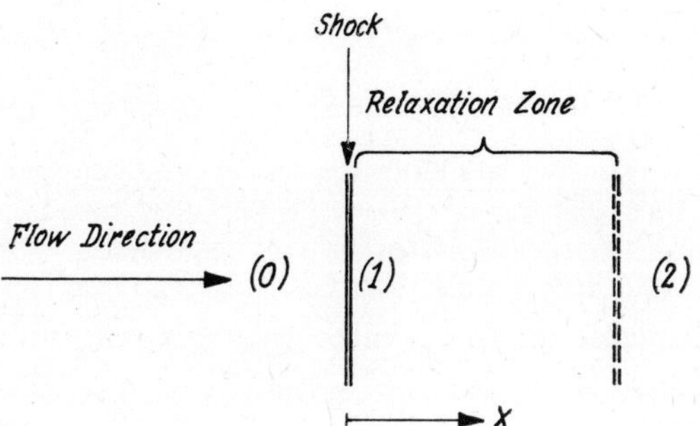

Fig. 4.1 New zone definition—compare Fig. 1.1.

The numbers (0) and (2) determine as before the homogeneous flow regions sufficiently far ahead of and behind the shock front where the chemical-thermodynamic equilibrium is established. An intermediate zone immediately behind the shock front is in nonequilibrium state.

The gas in this region goes through a continuous series of states starting with state (1) behind the shock front and finally reaching state (2) which, strictly speaking, is approached only asymptotically. Therefore, the extent of the relaxation zone is not sharply limited in the downstream direction. An extent of the relaxation zone can be defined, however, based on a certain convention. The proper shock front is assumed as a discontinuity surface in the mathematical sense, that means with infinitesimal small thickness.

For gases with relaxation, the ratio γ of the specific heat capacities can not be considered as constant. Some examples concerning the dependence of the average molar heat capacity at constant volume

$$\bar{C}_v = M_{mol}\bar{c}_v = M_{mol}\frac{E}{T} \tag{4.4}$$

on the number of atoms in the molecules at different temperatures are given in Table 4.1. According to Section 2.3 it was found that for perfect gases

$$C_v \approx f\frac{\text{cal}}{K \cdot \text{mol}}.$$

Table 4.1 Molar heat capacities $\bar{c}_v$ for perfect gases according to [23].

T [K]	O_2	N_2	CO_2	H_2O
0	2.98	2.98	2.98	2.98
300	4.96	4.96	5.52	5.96
500	5.06	4.98	6.46 (!)	6.07 (!)
1000	5.5	5.2	8.2	6.6
2000	6.1	5.75	10.1	7.8
3000	6.5	6.1	11.0	8.7
5000	7.07	6.46	11.9	9.8

The values listed above are valid for equilibrium regarding the energy forms *tr*, *rot*, *osc*, *exc*, excluding *chem* and *ion*. At $T = 0$ only the translational degrees of freedom are excited. At 300 K for O_2, N_2 and H_2O, only the energy forms *tr* and *rot* ($f = 5$ and 6 respectively) are excited, for CO_2 also *osc* to some degree. A crossing of the values of $\bar{c}_v$ which, according to the outlines in Sec. 2.3, is expected for the linear CO_2 molecule and the angular H_2O molecule, is indicated by the values of Table 4.1.

The corresponding change of γ may be explained in the case of the nitrogen molecule N_2:

Energy forms:	tr	rot	osc	diss	exc	ion
Quasi degrees of freedom	3	2	2			
$\gamma(N_2)$	7/5 (tr + rot)	9/7 (tr + rot + osc)				

To which extent the different degrees of freedom of a gas are excited depends on the temperature of the gas. In general a simultaneous partwise excitation of the different degrees of freedom is found*.

* An energy form not excited at low temperature equilibrium states is sometimes called "frozen". Here a different meaning of a "frozen state" is assumed which will be defined in the following.

If chemical reactions take place the molar heat capacity $\bar{C}_v$ or the formal number of the quasi degrees of freedom can be considerably increased.

An inert energy form *(vibr, chem, exc, ion)* which, after a rapid change of state as e.g. in a shock wave, leaves its amount of energy nearly unchanged is called *frozen*. According to the formalism used in this text this means $\xi_j = \text{const.}$ or $r_j = 0$.

The domain of relaxation phenomena ($0 < |r_j| < \infty$) is found between the two limiting cases of the shifting equilibrium ($r_j \to \infty$) and the frozen states ($r_j \to 0$). To indicate if a quantity is related to state changes with frozen degrees of freedom a subscript *f* or *fr* is used, and a subscript *e* or *eq* if the quantity refers to a shifting equilibrium process. For example,

$$\begin{aligned} \gamma_f &= \tfrac{7}{5} \quad \text{vibration frozen,} \\ \gamma_e &= \tfrac{9}{7} \quad \text{vibration in equilibrium.} \end{aligned} \tag{4.5}$$

A direct consequence resulting from Eq. (3.19) is the dependence of the velocity of sound on the change of state which is considered. For the two limiting cases of nonequilibrium with frozen energy forms and the shifting equilibrium the two different velocities of sound are found

$$\begin{aligned} a_f &= \sqrt{\gamma_f RT} \quad \text{(frozen)} \\ a_e &= \sqrt{\gamma_e RT} \quad \text{(equilibrium)} \end{aligned} \tag{4.6}$$

or according to the general definition of the velocity of sound

$$\begin{aligned} a_e^2 &= \left(\frac{\partial p}{\partial \varrho}\right)_{S|\xi_i=\tilde{\xi}_i} = -\frac{(\partial H^e/\partial \varrho)_p}{(\partial H^e/\partial p)_\varrho - 1/\varrho} \\ &= \left(\frac{\partial p}{\partial \varrho}\right)_{S,\xi_i} + \sum_{i=1}^{n}\left[\left(\frac{\partial p}{\partial \xi_i}\right)_{S,\varrho|\xi_i=\tilde{\xi}_i} \cdot \left(\frac{\partial \tilde{\xi}_i}{\partial \varrho}\right)_S\right], \end{aligned} \tag{4.7}$$

$$a_f^2 = \left(\frac{\partial p}{\partial \varrho}\right)_{S,\xi_i} = -\frac{(\partial H/\partial \varrho)_{p,\xi}}{(\partial H/\partial p)_{\varrho,\xi} - (1/\varrho)}. \tag{4.8}$$

For the derivation see e.g. [26] and [9] Ch. VIII.

It is always found that:

$$a_f \geq a_e; \tag{4.9}$$

a proof is given in Ref. [5] p. 45.

Vibrational relaxation in an acoustic wave e.g. is the result of nonequilibrium states. a_e and a_f are limiting values which are approached for sound frequencies $\omega \ll 1/\tau$ and $\omega \gg 1/\tau$ respectively (τ relaxation time, ω angular frequency).

The velocity of sound $a = a(\omega)$ for a matter which is at equilibrium in the undisturbed state is defined for any frequency ω. For the progress of a reaction where $\xi_i = \xi_i(t)$, only the velocity a_f is defined before and during the reaction process. $\xi_i = \tilde{\xi}_i$ is obtained only after the reaction has proceeded.

4.4 *Shock Waves*

We consider a gas described by the equation of state (4.2) which initially is in an equilibrium state (0) with

$$p_0, v_0, \xi_i^0 = \tilde{\xi}_i(p_0, v_0), \quad H_0 = H(p_0, v_0, \xi_i^0).$$

The state behind the relaxation zone where viscosity and thermal conduction may be neglected is described by the Hugoniot equation (3.11)

$$H(p, v, \xi_i) - H_0 = \tfrac{1}{2}(v + v_0)(p - p_0) \tag{4.10}$$

and the mechanical condition (3.8)

$$p - p_0 + (u_0 \varrho_0)^2 (v - v_0) = 0. \tag{4.11}$$

This should be explained in some more detail. The valid application of the Hugoniot equation (3.11) and the equation for the Rayleigh line (3.8) to the flow on both sides of a shock wave is essentially based on the assumption that outside the actual shock layer local transfer processes can be neglected. That means the terms regarding the influence of viscosity and thermal conduction in Eq. (1.2) and (1.3) must be negligible.

A reaction or relaxation induced by a shock wave and proceeding thereafter as an endothermic or exothermic process may, owing to

$$\left(\frac{du}{dx}\right)_{\text{react-zone}} \ll \left(\frac{du}{dx}\right)_{\text{shock}},$$

satisfy the shock conditions during the whole process. This is of course not true for the states within the shock transition zone.

Equation (4.11) is satisfied particularly for the so far unknown states in region (1), (see Fig. 4.1) where

$$p = p_1, \quad v = v_1, \quad \xi_i = \xi_i^0 \tag{4.12}$$

and region (2) where

$$p = p_2, \quad v = v_2, \quad \xi_i = \tilde{\xi}_i(p_2, v_2). \tag{4.13}$$

Introducing the values (4.12) and (4.13) respectively into Eq. (4.10) we obtain two different Hugoniot curves as shown in Fig. 4.2:

$\mathscr{H}_f$: For the actual, frozen shock;

$\mathscr{H}_e$: For the shock transition in the broader sense including relaxation until the equilibrium state is attained.

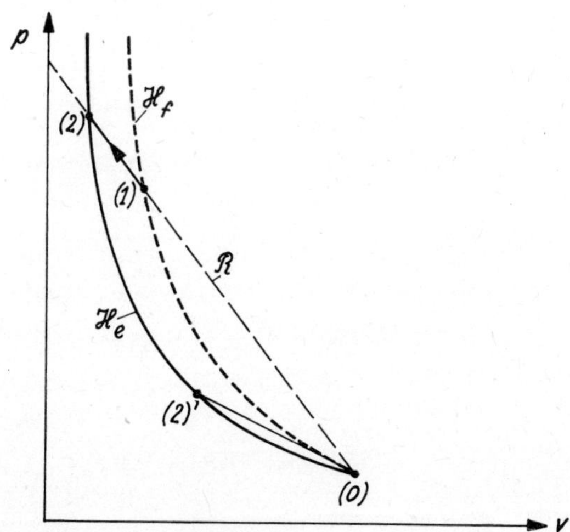

Fig. 4.2 Relaxation behind a shock wave.

The intersections of the curves $\mathscr{H}_f$ and $\mathscr{H}_e$ with their common Rayleigh line $\mathscr{R}$ according to (4.11) determine the two state points (1) and (2).

This behavior may be described phenomenologically as a shock transition in two different gases, a *frozen gas* with the equation of state

$$H^f(p, v) = H(p, v, \xi_i^0) \tag{4.14}$$

and an *equilibrium gas* (compare the discussion in Sec. 4.1) with the equation of state

$$H^e(p, v) = H(p, v, \tilde{\xi}_i(p, v)). \tag{4.15}$$

The relaxation process corresponding to a continuous transition from (1) to (2) is constrained during its progress to the Rayleigh line $\mathscr{R}$. The progress of the relaxation regarding the variables x, t, ξ_i is essentially determined by the reaction velocities (4.3). Its detailed description is not required for the present discussion.

For an energy-consuming relaxation, the point (2) on $\mathscr{R}$ lies, as will be seen later, always on the left, above point (1), see Fig. 4.2. In general, it follows from relation (4.9) that at point (0) the slope of the tangent on $\mathscr{H}_e$ must be less steep than the slope of the tangent to $\mathscr{H}_f$ in this point, because of the general relation

$$\left(\frac{\partial p}{\partial v}\right)_{\mathscr{H},0} = -a_0^2 v_0^2, \tag{4.16}$$

see also Eqs. (3.17), (3.18) and Fig. 3.5.

For a real shock wave with relaxation it is required according to (3.17) that

$$u_0 > (a_f)_0 . \tag{4.17}$$

A (0)–(2)′ transition, see Fig. 4.2, with

$$(a_f)_0 \geqq u_0 > (a_e)_0 \tag{4.18}$$

would correspond to a fully dispersed shock as discussed by LIGHTHILL in [24] p. 342, see also [5] p. 209. This special case represents a relaxation without an actual shock front.

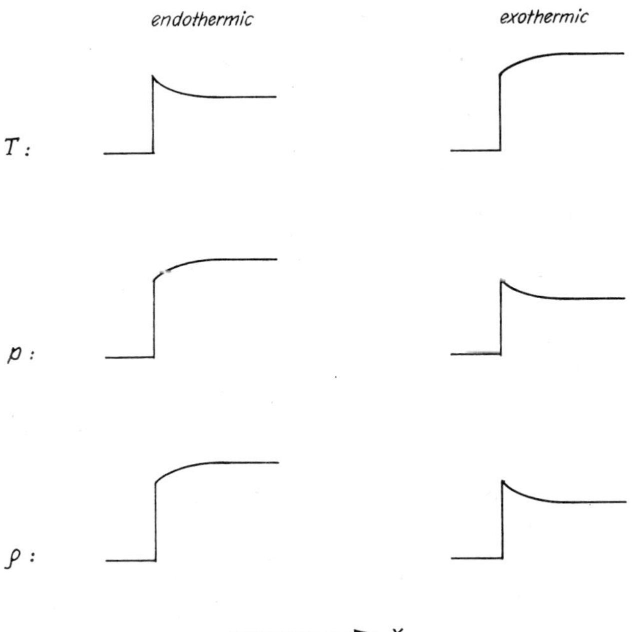

Fig. 4.3 Qualitative change of the state variables p, ϱ, T behind a shock wave with relaxation; *endothermal:* e.g., equilibrium adjustment of the vibrational degrees of freedom; *exothermal:* e.g., detonation, condensation.

For mechanical energy producing reactions (exothermal, expansive reactions; see Section 5) the process goes downward along $\mathscr{R}$ instead of upward and the equilibrium Hugoniot curve $\mathscr{H}_e$ no longer passes through point (0). In this case $\mathscr{H}_e$ runs always above point (0).

The fundamental qualitative development of the state variables p, ϱ, T for a shock wave with a subsequent endothermic relaxation of the degrees of freedom as well as an exothermic reaction is shown in Fig. 4.3.

The density changes in a gas flow across a shock wave can be made visible directly in an interferogram. Figure 4.4 shows an interferogram of a shock wave in CO_2 (reproduced from Ref. [24]). The fringe displacement is directly proportional to the density change. The curved fringes following the discontinuity therefore correspond directly to the density development in the relaxation zone. This corresponds to the ϱ curve shown for an endothermic process in Fig. 4.3.

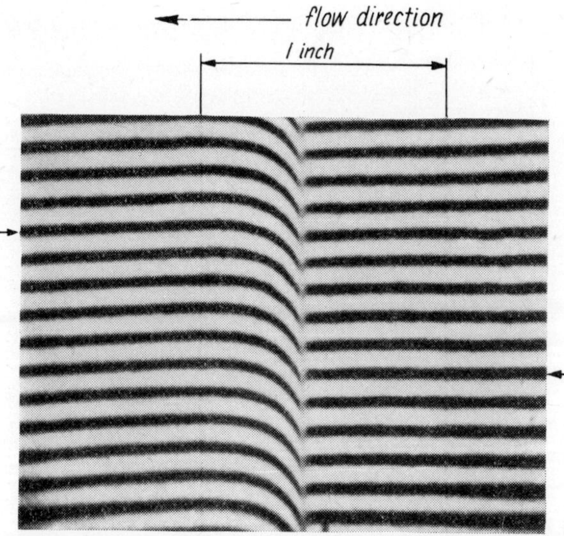

Fig. 4.4 Interferogram of a partly dispersed shock in CO_2 (Data: $T = 23\,°C, p = 260\,mb$, shock strength $\Delta p/p = 0.38$, total density increase 28%—corresponding fringes marked by arrows). From Ref. [24].

5 Reaction Energies

5.1 *Definitions*

A reaction which proceeds in a system causes a conversion of internal energy from one form into another.

Suppose the energy exchange with the environment in only due to volume work and heat transfer. The change in the internal energy is then given, according to (2.30), by

$$dE = \delta Q_e - p\,dv. \tag{5.1}$$

The heat added to the system is

$$\begin{aligned}\delta Q_e &= dE + p\,dv \\ &= dH - v\,dp.\end{aligned} \tag{5.2}$$

Therefore, for constant v and p, respectively, the differences in internal energy and enthalpy between the initial and final state are given by

$$\begin{aligned}(\Delta Q_e)_v &= (\Delta E)_v \\ (\Delta Q_e)_p &= (\Delta H)_p.\end{aligned} \tag{5.3}$$

This particularly applies for reaction processes.

For chemical reactions we primarily distinguish between three kinds of processes:

1) T, v *constant* (isothermal and isochoric)

$$Q_{T,v} = (\Delta E)_{T,v}, \tag{5.4}$$

2) T, p *constant* (isothermal and isobaric)

$$Q_{T,p} = (\Delta H)_{T,p}, \tag{5.5}$$

3) p, v *constant* (isobaric and isochoric*)

$$Q_{p,v} = (\Delta Q_e)_{p,v} = (\Delta E)_{p,v} = (\Delta H)_{p,v}. \tag{5.6}$$

For a perfect gas in cases 1) and 2) the quantities depend only on T. We therefore obtain

$$Q_{T,v} = (\Delta E)_T; \quad Q_{T,p} = (\Delta H)_T$$

and with $H = E + pv = E + nR_aT$

$$Q_{T,p} = Q_{T,v} + R_aT\,\Delta n. \tag{5.7}$$

Δn is the increase of the specific number of moles.

The heat $\Delta H = Q_{pT}$ which is to be added isobarically and isothermally for one formula reaction is tabulated for a great number of perfect gas reactions as a function of T; see e.g. Ref. [23].

* Compare Ref. [18] p. 7; [15].

The reaction process is for

$$\begin{aligned} Q_{T,v} \quad \text{or} \quad Q_{T,p} &< 0 \quad \text{exothermal,} \\ Q_{T,v} \quad \text{or} \quad Q_{T,p} &> 0 \quad \text{endothermal,} \\ Q_{p,v} &< 0 \quad \text{expansive,} \\ Q_{p,v} &> 0 \quad \text{contractive.} \end{aligned} \tag{5.8}$$

The quantities Q defined in this way are gross reaction energies. As negative values, they determine the heat which would be gained in a diabatic process under the given conditions.

Gasdynamic flow processes, however, proceed in general adiabatically with none of the quantities p, v, T being constant. Bulk quantities are then no more applicable and must be replaced by the corresponding differential quantities. From the equations of state we obtain the defining relations*:

$$E = E(v, T, \xi_i) \to q^j_{T,v} = \left(\frac{\partial E}{\partial \xi_j}\right)_{T,v}, \tag{5.9}$$

$$H = H(p, T, \xi_i) \to q^j_{T,p} = \left(\frac{\partial H}{\partial \xi_j}\right)_{T,p}, \tag{5.10}$$

$$H = H(p, v, \xi_i) = E + pv \to q^j_{p,v} = \left(\frac{\partial E}{\partial \xi_j}\right)_{p,v} = \left(\frac{\partial H}{\partial \xi_j}\right)_{p,v}. \tag{5.11}$$

For the quantities q, the same rule (5.8) is valid, describing the character of the reaction number j in the direction $d\xi_j/dt > 0$ (forward reaction)†.

For perfect gases, similarly as for Q, the relations hold

$$q^j_{T,v} = \left(\frac{\partial E}{\partial \xi_j}\right)_T,$$

$$q^j_{T,p} = \left(\frac{\partial (E + nR_aT)}{\partial \xi_j}\right)_T = q^j_{T,v} + R_aT\frac{\partial n}{\partial \xi_j}. \tag{5.12}$$

The gross quantities are given by integrals of the form

$$Q_{v,T} = \int q_{v,T}\, d\xi, \quad v, T = \text{const.} \tag{5.13}$$

---

* The differentiation $\partial/\partial\xi_j$ requires that all ξ_i with $i \neq j$ are kept constant.

† If the single reaction only is important for the description the suffix j is sometimes omitted.

The integration is to be taken over the interval from the initial to the final state. These limiting states may be fixed by definition, e.g. $\xi = 0$ and $\xi = 1$. For a perfect gas we then may find $q_{p,T} = \text{const} = Q_{p,T}$, $q_{v,T} = \text{const} = Q_{v,T}$, but in general $q_{p,v} \neq \text{const}$.

The bulk quantities $Q_{p,T}$, $Q_{v,T}$, and $Q_{p,v}$ which belong to three completely different final states cannot be related to each other by a general equation. For the corresponding differential quantities, however, such a general relation exists; see below Eq. (5.20).

Conceptionally for reaction energies we distinguish between:

1) Reaction energies in general for a certain reactive composition,

   a) absolute, measured e.g. in Joule or kcal,
   b) specific, measured e.g. in J/kg,
   c) molar or per formula reaction, measured e.g. in kcal/mole.

2) Reaction energies as a function of the reaction process,

   a) differential as $q = \partial Q/\partial \xi$,

   b) gross as $Q = \int_{\xi_2}^{\xi_1} q \, d\xi$.

3) Reaction energies

   a) for a reaction defined by a particular formula (e.g. written as: $H_2 + \frac{1}{2}O_2 + \Delta H = H_2O$),
   b) for equilibrium conditions (the equilibrium composition being in general dependent on the state).

4) Reaction energies for a prescribed progress of the reaction with only one free thermodynamic variable.

   a) $p, T = \text{const}$,
   b) $v, T = \text{const}$,
   c) $p, v = \text{const}$.

5) Reaction energies related

   a) to certain standard conditions (e.g. $p = 0$, $T = 0$, or $p = 1$ atm, $T = 0°C$),
   b) to the particular initial state (e.g. $p = p_0$, $v = v_0$).

5.2 *General relations for the differential quantity q*

Calculating the total differentials we obtain for the mechanical equation of state (4.1), (4.2)

$$dE = \left(\frac{\partial E}{\partial p}\right)_{v,\xi_i} dp + \left(\frac{\partial E}{\partial v}\right)_{p,\xi_i} dv + \sum_j \left(\frac{\partial E}{\partial \xi_j}\right)_{p,v} d\xi_j$$

or briefly

$$dE = \frac{\partial E}{\partial p}\, dp + \frac{\partial E}{\partial v}\, dv + \sum_j q^j_{p,v}\, d\xi_j. \tag{5.14}$$

The thermodynamic quantities $\partial E/\partial p$, $\partial E/\partial v$ have the same meanings as in (2.27), now being, however, "frozen quantities" due to $\xi_i = \text{const}$. Accordingly the following relations hold, see (4.8),

$$\Gamma_f = \frac{a_f^2}{pv} = \left(\frac{\partial \ln p}{\partial \ln \varrho}\right)_{S,\xi_i} = \left(\frac{\partial H}{\partial E}\right)_{S,\xi_i} = -\frac{v}{p}\,\frac{(\partial E/\partial v) + p}{(\partial E/\partial p)} = -\frac{v}{p}\,\frac{(\partial H/\partial v)}{(\partial E/\partial p)} \tag{5.15}$$

and also

$$(c_p)_f = \left(\frac{\partial H}{\partial T}\right)_{p,\xi_i}; \quad (c_v)_f = \left(\frac{\partial E}{\partial T}\right)_{v,\xi_i}, \tag{5.16}$$

$$\alpha_f = \frac{1}{v}\left(\frac{\partial v}{\partial T}\right)_{p,\xi_i}. \tag{5.17}$$

Between $q_{v,T}$, $q_{p,T}$, and $q_{p,v}$, according to [15], p. 16 the following general relation holds:

$$\gamma q_{v,T} = q_{p,T} + (\gamma - 1)\, q_{p,v}. \tag{5.18}$$

For the derivation of this relation we consider the following processes in the same system:

Reaction process	Heat added	Condition for the reaction process	Temperature increase
(1) $d\xi = r\,dt$	$q_{p,v}\,d\xi$	$p, v = \text{const}$	dT
(2) $d\xi = r\,dt$	$q_{p,T}\,d\xi$	$p, T = \text{const}$	0
(3) $d\xi = r\,dt$	$q_{v,T}\,d\xi$	$v, T = \text{const}$	0
(4) $d\xi = 0$	$(q_{p,v} - q_{p,T})\,d\xi$	$p, \xi = \text{const}$	dT
(5) $d\xi = 0$	$(q_{p,v} - q_{v,T})\,d\xi$	$v, \xi = \text{const}$	dT

The cases (4) and (5) are logically obtained by performing the differences: (4) = (1) – (2); (5) = (1) – (3). According to (4) and (5) we have,

$$\left.\begin{aligned}\frac{dT}{d\xi} &= \frac{q_{p,v} - q_{p,T}}{c_p} = \left(\frac{\partial T}{\partial \xi}\right)_{p,v},\\ \frac{dT}{d\xi} &= \frac{q_{p,v} - q_{v,T}}{c_v} = \left(\frac{\partial T}{\partial \xi}\right)_{p,v},\end{aligned}\right\} \tag{5.19}$$

and therefore

$$\frac{q_{p,v} - q_{p,T}}{c_p} = \frac{q_{p,v} - q_{v,T}}{c_v}, \tag{5.20}$$

$$q_{p,v} - q_{p,T} = \gamma\,(q_{p,v} - q_{v,T}), \tag{5.21}$$

or (5.18).

For reactions in a mixture of perfect gases without a change in the number of moles and at the absolute zero (temperature), according to (5.12) $q_{v,T} = q_{p,T}$ and therefore

$$q_{v,T} = q_{p,T} = q_{p,v} \quad \text{(for perfect gas, } \Delta n = 0 \text{ or } T = 0). \tag{5.22}$$

5.3 *The quantity* $q_{p,v}$

The quantity $q_{p,v}$ determines the *mechanical energy release* gained from reaction j,

$$q^i_{p,v} = q_j = \left(\frac{\partial E}{\partial \xi_j}\right)_{p,v} - \left(\frac{\partial H}{\partial \xi_j}\right)_{p,v}. \tag{5.23}$$

In order to explain the significance of this quantity for the description of a reaction process, we consider the thermal dilatation of a substance caused by a temperature increase at constant pressure which is determined by the thermal expansion coefficient $\alpha_f = (\partial \ln v/\partial T)_{p,\xi_i}$.

If α_f is divided by the specific heat capacity at constant pressure $(c_p)_f$, a mechanical quantity results:

$$\frac{\alpha_f}{c_{pf}} = \left(\frac{\partial \ln v}{\partial H}\right)_{p,\xi}. \tag{5.24}$$

The above quotient determines the dilatation of a non-reactive matter for isobaric heat addition.

The energy release of a reaction, on the other hand, is described by $q^j_{p,v}$. Therefore, for a reactive system the corresponding quantity (5.24) is given by

$$\sigma_j = -\frac{\alpha_f}{c_{pf}} q_j \tag{5.25}$$

$$= \left(\frac{\partial \ln v}{\partial \xi_j}\right)_{p,H} = \frac{v}{a_f^2}\left(\frac{\partial p}{\partial \xi_j}\right)_{v,U} \tag{5.26}$$

$$= \frac{1}{\Gamma_f}\left(\frac{\partial \ln p}{\partial \xi_j}\right)_{v,U}, \tag{5.27}$$

see Ref. [25], [15].

σ_j is a mechanical quantity (see Section 2) and a measure of the (partial) *expansivity* or the mechanical work performed by the particular reaction j. This is according to (5.8)

$$q_j < 0 \mathrel{\hat{=}} \sigma_j > 0 \quad \text{for expansive reactions,}$$

$$q_j > 0 \mathrel{\hat{=}} \sigma_j < 0 \quad \text{for contractive reactions.}$$

This rule is based on the assumption that α is always positive. Violations of this condition may be expected near phase limits, as e.g. for water according to its anomalous behavior near the freezing point, but not under the presently considered circumstances.

Since $\alpha = (vc_p/Ta^2)\,(\partial p/\partial S)_v$, the condition $\alpha > 0$ is equivalent to Bethe-Weyl's third condition (3.12c).

The total mechanical effect in the case of several simultaneous reaction processes is quantitatively described by

$$\left(\frac{\partial \ln v}{\partial t}\right)_{p,ad} = \sum_j \sigma_j r_j = \Sigma. \tag{5.28}$$

For further discussions see [15] and Section 8.

The coefficients σ_j form a vector and are related to the definition of the progress variables, whereas the scalar quantity $\Sigma = \Sigma(p, v, \xi_i)$ describes in an invariant manner the *total* reaction-produced *expansivity* in the state (p, v, ξ_i). For equilibrium $\Sigma = 0$, which is due to the fact that then all $r_i = 0$. However, $\Sigma = 0$ is also possible otherwise. That means nonequilibrium states may exist in a reaction-energetic neutral position, with neither expansive nor contractive behavior.

For a perfect gas with Eq. (2.16) the following relation results from (5.3), (5.12), and (5.18)

$$q_{p,v} = \left(\frac{\partial E}{\partial \xi}\right)_{T,v} - \frac{1}{\gamma - 1} R_a T \frac{dn}{d\xi}. \tag{5.29}$$

According to (5.29), the mechanical-energetic effect of the reaction, as far as it is due to the change of the molar number, essentially depends on the temperature level at which the reaction proceeds. An example, taken from Ref. [15], for the thermal dissociation of carbon tetrachloride into graphite and chlorine gas is shown in Table 5.1.

Table 5.1 Reaction energies in Joule/g for $CCl_4 \rightarrow C + 2Cl_2$.

T [K]	$Q_{p,T} = q_{p,T}$	$Q_{T,v} = q_{T,v}$	$q_{p,v}$ $\xi = 0 \cdots 1$
0	696	696	$696 \cdots 696$
1000	666	613	$-14 \cdots 362$
2000	624	516	$-803 \cdots 16$
3000	571	410	$-1630 \cdots -308$

6 A Relaxation Model

Suppose that, following a sudden change of state, e.g. behind a shock wave, the equilibrium is approached with a velocity proportional to the respective instantaneous deviation from the equilibrium state. In case of a single reaction variable ξ this process is then described by the relaxation equation, see e.g. [8], 305,

$$\frac{d\xi}{dt} = r = -\frac{1}{\tau}(\xi - \tilde{\xi}). \tag{6.1}$$

This gives for $\tilde{\xi}$ = const and τ = const

$$\xi - \tilde{\xi} \sim e^{-t/\tau}. \tag{6.2}$$

τ is called the *relaxation time* and represents the time required to decrease the distance from equilibrium by the factor $1/e$. If more than one reaction variable has to be regarded the factor $1/\tau$ in (6.1) is to be replaced by a tensor.

This description of the equilibrium approach is always approximately valid if the deviations from the equilibrium state are small. We mention as an example the relaxation of the vibrational degrees of freedom of the molecules in an acoustic wave (with small deviations from the equilibrium).

In the general case r and τ are functions of p, v and ξ_i, and $\tilde{\xi}_i$ are functions of p and v. For chemical reactions, the influence of an induction period which is not taken into account by (6.1) very often has to be regarded. The quantity $\tau = \tau(p, v, \xi)$, defined by (6.1), may considerably vary in this case and (6.2) is not applicable for larger deviations from the equilibrium. This is particularly true for combustions and detonations.

For the diatomic nitrogen molecule N_2 e.g. the corresponding quantities have the following values (see Section 4.3)

$$\left.\begin{aligned} \gamma_f &= \tfrac{7}{5}; \quad f_{fr} = 5, \\ \gamma_e &= \tfrac{9}{7}; \quad f_{eq} = 7, \end{aligned}\right\} \tag{6.3}$$

$$E(p, v, \xi) = \tfrac{5}{2} pv + \xi, \tag{6.4}$$

$$\tilde{\xi} = pv, \tag{6.5}$$

$$E(p, v, \tilde{\xi}) = \tfrac{7}{2} pv. \tag{6.6}$$

In (6.4) the vibrational energy itself was taken as a progress variable. This is permissible since this energy contribution remains constant due to *freezing*. In (6.5) for simplicity, the vibrational energy is assumed as fully excited. Actually, $\tilde{\xi}$ is a monotonic, nonlinear function of p and v which may be obtained from tables for $H(T)$ according to

$$H\left(\frac{pv}{R}\right) = \frac{7}{2} pv + \tilde{\xi}\left(\frac{pv}{R}\right).$$

Resulting from Section 5 it is

$$\begin{aligned} q_{p,v} &= 1, \\ \left(\frac{\alpha}{c_p}\right)_f &= \frac{2}{7pv}; \quad (c_p)_f = \frac{7}{2} R, \\ \sigma &= -\frac{2}{7pv}, \end{aligned} \tag{6.7}$$

that means the reaction proceeds as contractive for increasing ξ. The quantity Σ becomes according to (5.28) and (6.1), (6.7)

$$\Sigma = r\sigma = \frac{\xi - \tilde{\xi}}{\tau} \frac{2}{7pv} = \frac{\xi/pv - 1}{3{,}5\,\tau}. \tag{6.8}$$

Figure 6.1 (see also Fig. 4.2) shows the Hugoniot curves $\mathscr{H}_f$ and $\mathscr{H}_e$ in a dimensionless representation for the present case. Equilibrium exists in the initial state,

$$\xi_0 = p_0 v_0.$$

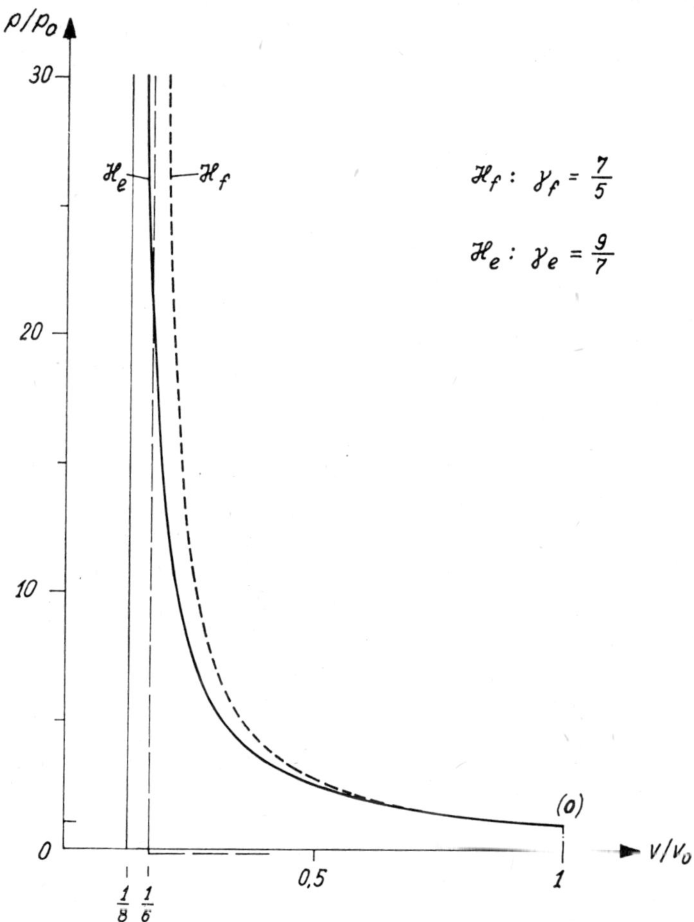

Fig. 6.1 Hugoniot curves for the described relaxation model.

The equations for the $\mathscr{H}$-curves are according to (3.22) and with the above values for γ_f and γ_e

$$\mathscr{H}_f\colon \quad \frac{v}{v_0} = \frac{p/p_0 + 6}{6p/p_0 + 1}, \tag{6.9}$$

$$\mathscr{H}_e\colon \quad \frac{v}{v_0} = \frac{p/p_0 + 8}{8p/p_0 + 1}. \tag{6.10}$$

Behind the frozen shock, that means on $\mathscr{H}_f$ above point (0),

$$\xi_1 = \xi_0 = p_0 v_0 < p_1 v_1 = \tilde{\xi}_1.$$

(6.8) shows that $\Sigma_1 < 0$, and then the larger $|\Sigma_1|$, the smaller $p_0 v_0 / p_1 v_1 = T_0 / T_1$.

While the state during the relaxation process is changing along $\mathscr{R}$ from (1) to (2) (see Fig. 4.2) $\tilde{\xi}$ changes its value from $p_1 v_1$ into $p_2 v_2$. Therefore relation (6.2) does not hold, or at least not exactly.

The differential equations for the change of state will be derived in Section 8, and the relaxation model will be discussed again with a slight modification in Section 14.

A more complicated and more general model with two reaction variables is discussed in [26]. It is outlined there under which conditions the relaxation process proceeds qualitatively as discussed above.

CHAPTER III

Combustion, Detonation

7 The Hugoniot Curve for Combustion and Detonation Processes

7.1 *Detonation*

The discussion of a *detonation front* propagating in a gas will be based on the following idea.

A shock wave propagates into a reactive gas mixture which is in a metastable thermodynamic-chemical (pseudo-) equilibrium with frozen reaction. The intensity of the shock wave and the corresponding change of state is sufficiently large to start the reaction process. In other words, we consider the detonation as a shock-induced-reaction front*.

It is thereby assumed that the detonation is a self-sustaining process, at least for a certain time interval. That means the reaction is assumed to release sufficient energy to drive the shock wave which induces the reaction.

If the shock intensity does not depend on the reaction process we shall talk about a *forced detonation* or a *shock-induced combustion*.

The detonation process described in a p–v diagram is shown in Fig. 7.1.

Two different Hugoniot curves are determined by the initial state where the gas mixture or the explosive is in metastable equilibrium with frozen reaction and the final state determined by the reaction products being at equilibrium

$$\mathscr{H}_0: \quad \xi_i = \xi_i^0,$$

$$\mathscr{H}_e: \quad \xi_i = \tilde{\xi}_i(p, v).$$

It is assumed that the reaction is described by a number ($i = 1, \ldots, n$) of parameters ξ_i with the initial and final values ξ_i^0 and $\tilde{\xi}_i$. See also the discussion of shock waves with relaxation in Section 4.

* This model is due to J.B. ZELDOVICH, J. v. NEUMANN and W. DÖRING and is briefly called the ZND model [32] [29].

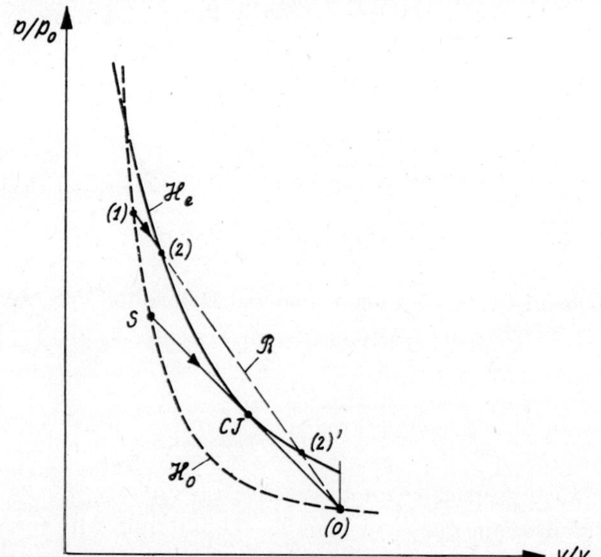

Fig. 7.1 Detonation branch of the Hugoniot curve.

Now, if the propagation velocity of the detonation front is known, then consequently the slope of the Rayleigh line $\mathcal{R}$ is given, and the points of intersection of $\mathcal{R}$ with the curves $\mathcal{H}_0$ and $\mathcal{H}_e$ determine the states immediately behind the shock front and after the reaction process has been completed.

Suppose at first that the shock wave starting the reaction may be prescribed for the system by some external source. That means the propagation velocity or intensity of the shock wave is given independently of the reaction process (forced detonation). Then the states taken by the gas will be in the following order, see Fig. 7.1:

a) Initial state (0).
b) State (1) immediately behind the shock front.
c) Equilibrium state (2), after the reaction has proceeded.

The pressure peak which, according to this behavior, occurs behind the shock front (point (1) or S in Fig. 7.1) is often called the *von Neumann spike.*

State (2)′ can in general not be obtained with the above order of changes of state. A discontinuous transition (2) → (2)′ would correspond to a rarefaction discontinuity behind the reaction zone in the gas containing the reaction products; this was excluded in 3.6. A change of state along the Rayleigh line from (2) → (2)′ on the other hand would be possible only by heat addition up to the sonic limit with subsequent cooling; see the discussion in Section 8.

The progress of the reaction within the relaxation zone from (1) → (2) is constrained to the Rayleigh line if viscosity effects can be neglected.

There exist obviously two limiting positions for the Rayleigh line. The first one is obtained when $\mathscr{R}$ is parallel to the p-axis:

$$v = v_0 = \text{const}; \quad \theta^2 = u_0^2 \to \infty; \tag{7.1}$$

this would correspond to an infinite flow velocity u_0 in front of the detonation wave (or to a detonation with constant specific volume).

The other limiting position is obtained when the Rayleigh line becomes a tangent to the Hugoniot curve $\mathscr{H}_e$ of the reaction products. The contact point is called the *CJ point**. We shall see later that behind the reaction zone at the point CJ the flow velocity is just equal to the velocity of sound.

The possible final states along the equilibrium Hugoniot curve $\mathscr{H}_e$ for the interval $v < v_0$, $p > p_0$ belong, due to the supersonic propagation velocity of the reaction front, obviously to the class of detonation phenomena. This upper branch of the Hugoniot curve is therefore called the *detonation branch.* The processes which lead to the various final states on $\mathscr{H}_e$, however, are different. Therefore, the following sections are usually distinguished on the detonation branch of the equilibrium Hugoniot curve $\mathscr{H}_e$:

Section CJ–(2): This is the region of the forced, overdriven, or *strong detonation.* The propagation velocity $D = u_0$ of the detonation front increases in the direction from CJ to (2). The flow velocity u_2 behind the detonation wave is smaller than the velocity of sound,

$$D > D_{\mathrm{CJ}}; \quad u_2 < a_2. \tag{7.2}$$

CJ Point: This particular final state corresponds to the so-called *Chapman–Jouguet Detonation.* The propagation velocity of the detonation front has reached its smallest possible value

$$D_{\mathrm{CJ}} = D_{\mathrm{Min}}; \quad u_2 = a_2, \tag{7.3}$$

when $\theta^2 = \min$.

---

* Chapman–Jouguet point.

Section CJ–(2)′: This is the region of the *weak detonation*. The propagation velocity of the reaction front has the same magnitude as for the corresponding strong detonation. The flow velocity behind the reaction zone, however, is supersonic

$$D > D_{CJ}; \quad u_2 > a_2. \tag{7.4}$$

Which one of the three types of detonations actually occurs will be discussed in Section 12 where the conditions for the stability of a detonation front are described.

If, as assumed in Fig. 7.1, an intersection between $\mathscr{H}_e$ and $\mathscr{H}_0$ exists at the strong detonation branch the usual distinction between the three cases as made above is no longer quite correct. For sufficiently large values of θ the Rayleigh line $\mathscr{R}$ will then be above this intersection point. Consequently, the reaction will proceed from (1) to (2) upward instead of downward and will therefore be contractive and not expansive. The entire wave is to be considered more as a shock wave with relaxation than as a strong detonation. It has been proved in [30] that such a point of intersection exists for any gas detonation resulting in an upper limit for the strong detonation region. It has not yet been proved that the same general behavior may also be true for condensed explosives though it seems to be very likely. Some diagrams in [31] seem to show that for nitro-methane and TNT this point of intersection exists.

7.2 *Combustion*

If the Rayleigh line $\mathscr{R}$ originating from point (0) on $\mathscr{H}_0$ does not meet the upper branch of the Hugoniot curve $\mathscr{H}_e$, then the equilibrium of the reactive gas mixture can not be approached by means of a shock transition.

In the case where the Rayleigh line intersects with the lower branch of the Hugoniot curve $\mathscr{H}_e$ with $v > v_0$ and $p < p_0$, a further possible final state is obtained by the corresponding process. Such processes which lead to a final state on the lower branch of the $\mathscr{H}_e$ curve are characterized by a subsonic propagation velocity of the reaction front and a decrease of the pressure and the density across the front. Processes of this kind are called *combustions*. Compare the diagram shown in Fig. 7.2.

A more appropriate notation would be the word *deflagration* to distinguish the flame front propagating in a homogeneous medium from the general combustion phenomena like e.g. diffusion flames. Furthermore, only an oxydation should be called combustion, while detonations and deflagrations may in principle be caused by pure disintegration of a substance.

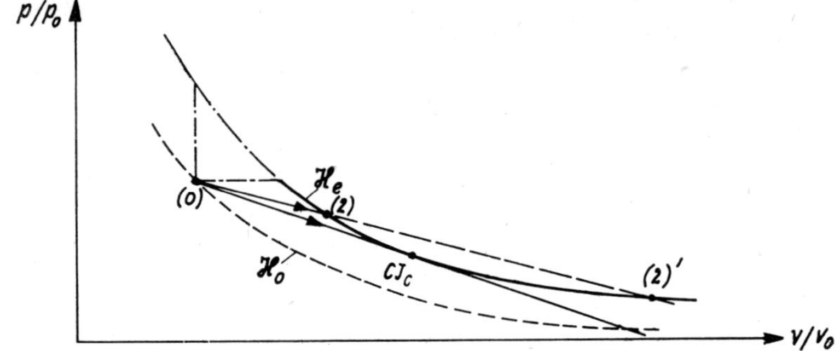

Fig. 7.2 Combustion (or deflagration) branch of the Hugoniot curve.

Since the tangent on $\mathscr{H}_0$ at point (0), for which $u_0 = a_0$, does not intersect the $\mathscr{H}_e$ curve, there is an interval of propagation velocities near $u_0 \approx a_0$ for which no steady, equilibrium approaching, reaction front exists.

Similarly as for a detonation we distinguish between a *strong* and *weak combustion* as well as a *Chapman–Jouguet combustion*, see Fig. 7.2, with the following relationship:

Final state (2): *Weak combustion*, with $u_2 < a_2$.
Final state CJ_c: *Chapman–Jouguet combustion*, with $u_{J_c} = a_{J_c}$.
Final state (2)′: *Strong combustion*, with $u_2' > a_2'$.

The fundamental differences between detonation and combustion are shown again in Figs. 7.3 and 7.4. For the description of the undisturbed flow in front of the reaction wave the Mach numbers

$$M = \frac{u}{a} \quad \text{(steady)} \tag{7.5}$$

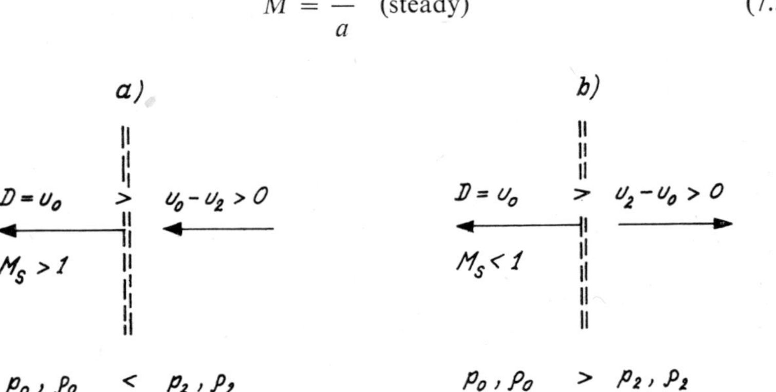

Fig. 7.3 Detonation (a) and combustion (b) in a fixed frame of reference. u_0 propagation velocity of the reaction front; $u_0 - u_2$ flow velocity of the reaction products.

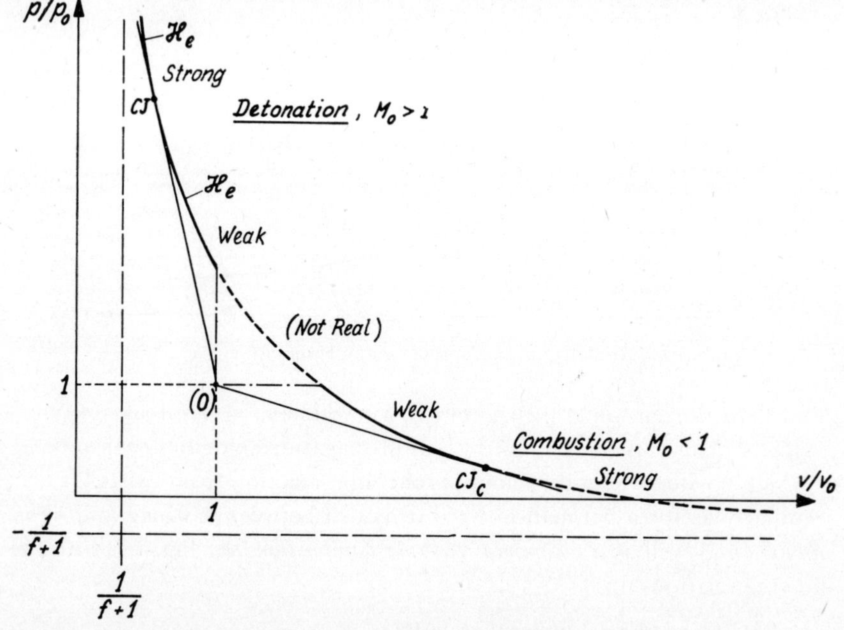

Fig. 7.4 Equilibrium Hugoniot curve for detonation and combustion.

and

$$M_s = \frac{D}{a_0} \quad \text{(unsteady)} \tag{7.6}$$

are used respectively.

7.3 *Classification of steady plane reaction fronts (Jouguet's rule)*

A systematic classification of the different kinds of reaction fronts leads to the following six types of waves *(Jouguet's rule)*. With respect to the reaction front, the flow is:

supersonic in front of a *detonation wave*, with supersonic flow behind a weak detonation, and subsonic flow behind a strong detonation;

subsonic in front of a *combustion wave*, with subsonic flow behind a weak combustion and supersonic flow behind a strong combustion;

sonic behind a *Chapman–Jouguet detonation* as well as behind a *Chapman–Jouguet combustion*.

The same six cases are listed in the following scheme:

		$M_{0,f}$	M_{2e}	Shock	Real?
Detonation	strong	> 1	< 1	yes	yes
	CJ	> 1	$= 1$	yes	yes
	weak	> 1	> 1	no (?)	yes
Combustion	weak	< 1	< 1	no	yes
	CJ	< 1	$= 1$	no	no
	strong	< 1	> 1	(Rarefaction discontinuity)	no

These results are valid under fairly general assumptions about the behavior of the Hugoniot curves; compare the conditions (3.12) according to BETHE and WEYL.

The information about the physical reality of the process given in the last column of the above scheme will be confirmed later.

7.4 *Entropy change along the equilibrium Hugoniot curve*

The area rule (3.13), relating the area swept over by the $\mathscr{R}$ line and the entropy change along $\mathscr{H}$, is also valid for detonation and combustion fronts. The proof did not contain the assumption that the initial point (0) was on $\mathscr{H}$.

Suppose the final state varies along $\mathscr{H}_e$ in the downward direction starting at $p = \infty$. The $\mathscr{R}$ line then in general turns at first to the left and finally, when $p \to 0$ once again to the left. Between these limiting positions it changes twice its rotating direction, namely at the tangential CJ detonation and CJ–combustion points. For final states on $\mathscr{H}_e$, therefore, the entropy S goes through a minimum at the CJ detonation point and a maximum at the CJ–combustion point. As for the rest S varies monotonically along $\mathscr{H}_e$.

The above discussion also pertains to the excluded interval on $\mathscr{H}_e$ where $\Delta p/\Delta v > 0$. The points along $\mathscr{H}_e$ for this interval represent realistic thermodynamic states which satisfy the Hugoniot equation; however, they cannot be obtained by steady reaction fronts with an initial state (0).

We therefore obtain the result that for any detonation wave the entropy change will be positive with a minimal increase for the final state CJ. Consequently, the curve $\mathscr{H}_e$ cannot intersect the isentrope $\mathscr{I}_e$ at the Chapman–Jouguet point CJ. Hence it follows that at the point CJ the Hugoniot curve $\mathscr{H}_e$, the Rayleigh line $\mathscr{R}$ and the corresponding isentrope $\mathscr{I}_e$ have the same slope *(contact theorem)*. See also Fig. 7.5.

For the CJ point on the combustion branch on the contrary, the entropy increase is maximal; see Fig. 7.5. Here too $\mathscr{R}$ is tangent to $\mathscr{H}_e$ and $\mathscr{J}_e$.

Since, owing to condition (3.12b), the curvature of the isentrope does not change its sign, each $\mathscr{R}$ line is tangent to the given field of isentropes in just one point (see Fig. 7.6). Compared with the local isentrope direction at

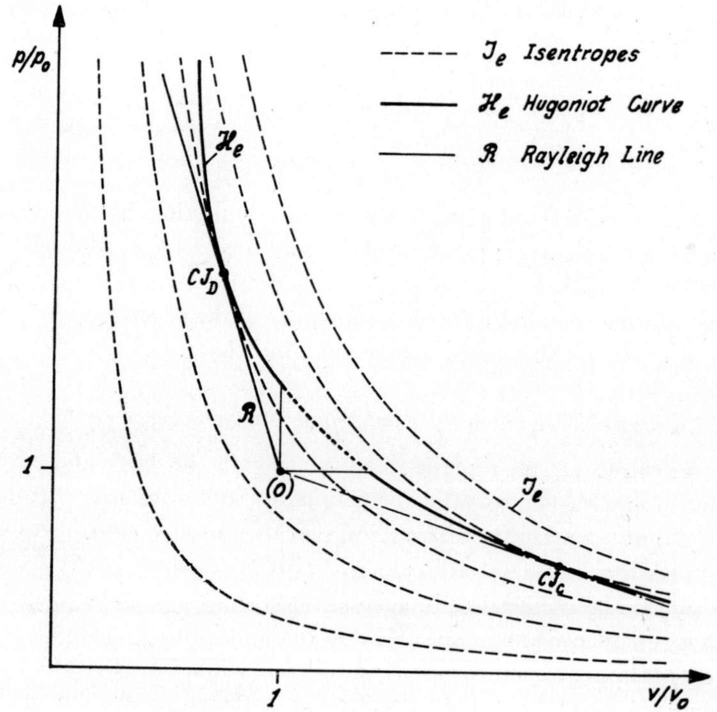

Fig. 7.5 The contact theorem.

the tangential point the slope of $\mathscr{R}$ is steeper below and flatter above this point. We now have according to (2.26)—see also (3.12a), (3.8) and (3.1)—,

$$\left(\frac{\partial p}{\partial v}\right)_S = -\frac{a^2}{v^2}, \quad \left(\frac{\Delta p}{\Delta v}\right)_{\mathscr{R}} = -\varrho^2 u^2. \tag{7.7}$$

Consequently, the quotient of the two slopes determines the local Mach number:

$$\frac{(\Delta p/\Delta v)_{\mathscr{R}}}{(\partial p/\partial v)_S} = M^2. \tag{7.8}$$

To every initial point (0) in the field of isentropes therefore belongs a certain *sonic line* or *sonic limit* with $M = 1$, see Fig. 7.6. This line is the geometrical locus of all tangential points. Along $\mathscr{R}$ there is always $M < 1$ above and $M > 1$ below the sonic point. This is valid for $p > p_0$ (detonation) as well as for $p < p_0$ (combustion). The limiting lines

$$v = v_0, \quad M = \infty$$

$$p = p_0, \quad M = 0$$

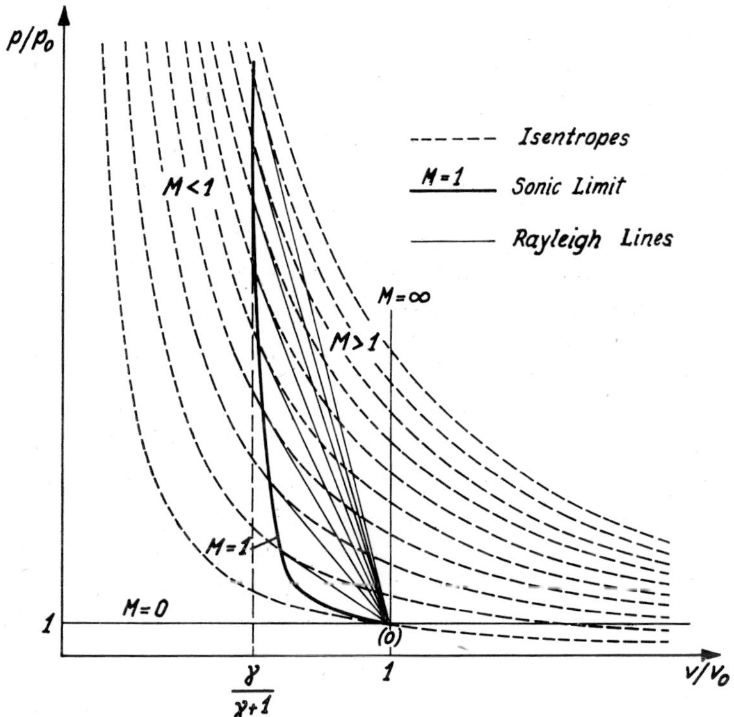

Fig. 7.6 The sonic limit for calorically perfect gas ($\gamma = 1.4$).

separate the detonation quadrant ($p > p_0$, $v < v_0$) and the combustion quadrant ($p < p_0$, $v > v_0$) from the two quadrants excluded for $\mathscr{R}$ lines ($\Delta p/\Delta v > 0$). The sonic line goes through (0) and is tangent to the isentrope at this point. In the case of a perfect gas the sonic line is an equilateral hyperbola

$$\left(p - \frac{p_0}{\gamma + 1}\right)\left(v - \frac{\gamma v_0}{\gamma + 1}\right) = \frac{\gamma p_0 v_0}{(\gamma + 1)^2}, \tag{7.9}$$

with its second branch passing through the point $p = 0$, $v = 0$.

Referring back to Fig. 7.5 we conclude that the $\mathscr{H}_e$ curve is intersected (not touched) by the equilibrium-sonic line at the two CJ points. Therefore the detonation branch as well as the combustion branch of $\mathscr{H}_e$ are divided into an upper subsonic and a lower supersonic section. This was pointed out before, though without proof.

7.5 *Related processes (condensation discontinuities)*

The discontinuous condensation which may occur in the supersonic flow of water vapor or moist air is accompanied by a heat release and satisfies a Hugoniot relation. It is a process akin to the detonation.

According to a recent contribution [37] two cases are to be distinguished in a supersonic flow with condensation:

a) Supersonic flow of a gas-vapor mixture. The heat release due to condensation leads to a thermal shock. The flow is supersonic throughout [34], [35], [36].

b) Supersonic flow of an undercooled vapor. A shock wave in the undercooled vapor causes condensation. The resulting heat release leads to a stabilization of the shock front *(vapor detonation)*.

The basic difference of the two processes a) and b) is explained best by means of their description in a p–v diagram, see Fig. 7.1. In case a) the states (0) and (2)′ are successively obtained, with a generally very fast change of state (0) → (2)′ similar to a shock transition. See also the outlines in Section 8.4, supersonic combustion.

In case b) for stable conditions the changes of state occur in the order (0) → S → CJ. This process corresponds to a regular detonation. An experimental verification of this phenomenon besides [37] apparently does not exist. See also [3] F. 4, and Section 8.

8 **Rayleigh Processes**

8.1 *Definitions*

From the balance equations (1.1) and (1.2) for the mass flow and the momentum flow the equation of the Rayleigh line in a p–v diagram was obtained by eliminating the flow velocity u_2

$$\frac{\Delta p}{\Delta v} = -\theta^2. \tag{8.1}$$

Flow processes which are governed by a linear p–v relation of the form (8.1) with constant θ will be called *Rayleigh processes.* With

$$\theta = \varrho_0 u_0 = \varrho u = \text{const} \tag{8.2}$$

the mass flow density is constant.

The restricting conditions for this type of flow are:

a) Constant cross-sectional area,
b) steady flow,
c) no viscosity,

whereas an increase of heat caused by reactions or external heat addition is possible (see [3] F. 4).

In the following the most important characteristics of this type of flow are listed as related to the present discussion.

8.2 *Change of state along the Rayleigh line*

For the reaction j, with $H = H(p, v, \xi_j)$, the following relations hold:

$$dH = \left(\frac{\partial H}{\partial p}\right)_{v,\xi_j} dp + \left(\frac{\partial H}{\partial v}\right)_{p,\xi_j} dv + \sum_j \left(\frac{\partial H}{\partial \xi_j}\right)_{p,v} d\xi_j, \tag{8.3}$$

and regarding (4.8), (5.16), (5.17) and (5.11)

$$\left(\frac{\partial H}{\partial p}\right)_{v,\xi_j} = v + \frac{c_{pf} v}{\alpha_f a_f^2}, \tag{8.4}$$

$$\left(\frac{\partial H}{\partial v}\right)_{p,\xi_j} = \left(\frac{c_p}{\alpha v}\right)_f = \frac{c_{pf}}{a_f v}, \tag{8.5}$$

$$\left(\frac{\partial H}{\partial \xi_j}\right)_{p,v} = q_j,$$

$$dH = \left(v + \frac{c_p v}{\alpha a^2}\right)_f dp + \left(\frac{c_p}{\alpha v}\right)_f dv + \sum_j q_j \, d\xi_j.$$

On the other hand, with (5.25):

$$dH - v\,dp = dE + p\,dv = \delta Q_e = 0,$$

$$0 = \frac{\alpha_f}{c_{pf}} (dH - v dp) = \frac{dv}{v} + \frac{v}{a_f^2} dp - \sum_j \sigma_j \, d\xi_j.$$

From Eq. (4.3) and the equation of the Rayleigh line (8.1) it follows that

$$\frac{dv}{v}\left(1+\frac{v^2}{a_f^2}\frac{dp}{dv}\right)-\left(\sum_j \sigma_j r_j\right)dt=0,$$

$$\frac{dv}{v}(1-M_f^2)-dt\cdot\Sigma=0$$

or

$$\boxed{\frac{dv}{dt}=\frac{v\cdot\Sigma}{1-M_f^2}.} \tag{8.6}$$

The differential equation (8.6) describes the change of the specific volume for an adiabatic* Rayleigh process with chemical reactions. Equation (8.6) in connection with (1.1), (4.3), (8.1) and

$$\frac{dx}{dt}=u \tag{8.7}$$

determines the variation of the variables x, t, ξ_j, u, p, v along $\mathscr{R}$ and thereby the complete structure of the reaction zone.

8.3 *Temperature and entropy change along the Rayleigh line*

A change of state along the Rayleigh line is in general accompanied by a temperature change as well as an entropy change. In the case of a perfect gas, a simple analysis using Eq. (8.1) for the $\mathscr{R}$ line, equation (2.16) for the isothermal curves and equation (2.39) for the isentrope shows that the following maximal values are obtained as

$$\left.\begin{aligned} &S_{\max}\ \text{for}\ M=1;\quad v=\frac{\gamma}{\gamma+1}v_m;\quad p=\frac{1}{\gamma+1}p_r\\ &T_{\max}\ \text{for}\ M=\gamma^{-1/2};\quad v=\frac{v_m}{2};\quad p=\frac{p_r}{2}. \end{aligned}\right\} \tag{8.8}$$

See Fig. 8.1.

According to (1.2), (3.2), and (3.5) $p_r = p_0 + \theta^2 v_0$ and $v_m = p_r/\theta^2$ are the extremal values on $\mathscr{R}$ (at $v = 0$ and $p = 0$, respectively).

* For the definition of adiabatic processes used in this text, see section 2.

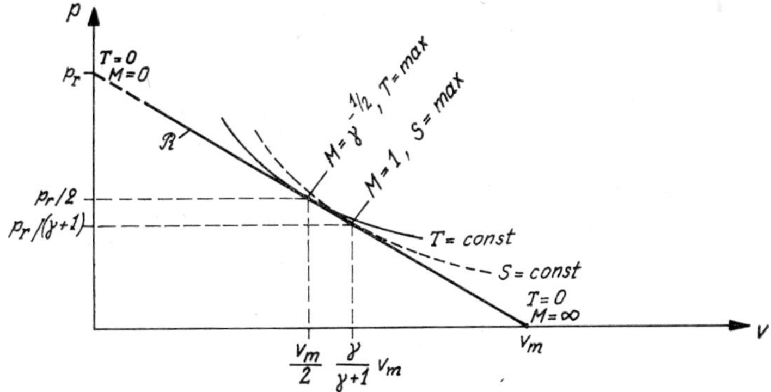

Fig. 8.1 Rayleigh process for a calorically perfect gas ($\gamma = 3/2$).

A Rayleigh process leading from $M < 1$ to $M > 1$ is possible with heat addition up to $M = 1$ and subsequent cooling (heat nozzle). Since

$$M^2 = \frac{u^2}{\gamma p v} = \frac{\theta^2 v}{\gamma p} \tag{8.9}$$

M varies monotonically along $\mathcal{R}$.

S_{max} is obtained for $M = 1$ in general, and also for real gases*, while relation (8.8) and Fig. 8.1 are only qualitatively valid in this case.

If $a_e \neq a_f$ and nonequilibrium states are obtained along $\mathcal{R}$, a_e and M_e do not exist there. S_{max} is again assumed at $M_f = 1$. However, for the particular medium in general no uniquely determined frozen isentrope field exists. But for any real state point on each $\mathcal{R}$ line a certain frozen-isentrope direction and a Mach number M_f can be found. The f-sonic limit is a curve similar to the e-sonic limit, but it generally occurs in the e-subsonic region.

For the relaxation model discussed in Section 6 the two sonic limits can be obtained from (6.3) and (7.9) as hyperbolas intersecting at point (0).

Further relations describing the change of state along the Rayleigh line are given in Ref. [33].

* In so far as the entropy change is due to an external heat transfer or to expansive reactions; however, cf. p. 87.

8.4 *Supersonic combustion*

For supersonic inlet conditions of an airbreathing jet engine, for example, the heat addition by combustion is performed after the flow has been decelerated by a shock wave to subsonic flow (supersonic diffusor). The heat increase may then accelerate the flow at most up to $M = 1$ supposing the cross-sectional area remains constant. Compare the order of changes of state $(0) \to (1) \to (2)$ in Fig. 7.1 and Fig. 9.2.

For the same supersonic-inlet flow, on the other hand, a Rayleigh process may be constructed which continuously, without shock, leads from state (0) to (2′), Fig. 7.1. This process would represent a true supersonic combustion with Mach numbers $M > 1$ throughout. It would correspond to the earlier discussed condensation discontinuity or thermal shock and is like the latter only possible as non-self-sustaining reaction front. A satisfactory experimental and technical performance of this supersonic combustion has not yet been realized. It would, however, be very desirable since for high inlet Mach numbers corresponding to steep $\mathscr{R}$ lines the process $(0) \to (1) \to (2)$ produces little or no energy (intersection of $\mathscr{H}_0$ and $\mathscr{H}_e$, transition to contractive reactions). The reaction (1) to (2′) on the contrary proceeds as an expansive process.

CHAPTER IV

Ideal Detonation of Perfect Gases

9 The Detonation Discussed by Means of a Model

A FAIRLY COMPREHENSIVE formal description of the detonation in a gas mixture can be made using a simple model; see e.g. [3] G, F4. This model makes several restrictive assumptions which of course are not met in the general real case.

9.1 *The model*

The first basic assumption is that in the mechanical equation of state (4.1) a formal separation between the solely thermal and the solely chemical processes is possible.

A further restriction is that the unreacted gas mixture as well as the reaction products are assumed to be calorically perfect gases. Changes which may occur in the adiabat exponent are neglected.

This leads to the following relations,

$$E = E(p, v, \xi) = E_{th}(p, v) + E_{ch}(\xi) = \frac{pv}{\gamma - 1} + E_{ch}(\xi) \tag{9.1}$$

$$H = H_{th}(p, v) + E_{ch}(\xi) = \frac{\gamma pv}{\gamma - 1} + E_{ch}(\xi) \tag{9.2}$$

and with (2.21)

$$\gamma = \frac{C_p}{C_v} = \text{const}; \quad a_e = a_f.$$

For $E_{ch}(\xi)$ the particular dependence is assumed

$$E_{ch}(\xi) = (1 - \xi)\, Q_0, \tag{9.3}$$

with

$$\begin{aligned} &\xi = \xi_0 = 0 &&\text{for metastable equilibrium} \\ &\xi = \tilde{\xi} = 1 &&\text{for equilibrium,} \\ &\xi = \text{const} &&\text{for frozen states.} \end{aligned}$$

Let

$$Q = Q_0 - E_{ch}(\xi) \geq 0$$

be the chemical energy released between the initial state (0) and some intermediate state (p, v, ξ). We write

$$\frac{Q}{H_{0\,th}} = \frac{Q}{c_p T_0} = \frac{\gamma - 1}{\gamma} \frac{Q}{p_0 v_0} = \bar{Q}. \tag{9.4}$$

The quantity $\bar{Q}$ is due to G. DAMKÖHLER [38] and may be called the *Damköhler parameter**. $\bar{Q}$ represents a dimensionless reaction energy and depends according to our assumption only on the reaction variable ξ.

Regarding the assumption made with Eq. (9.1) when establishing the Hugoniot equation,

$$\frac{1}{\gamma - 1}(pv - p_0 v_0) - Q = \Delta E = -\frac{1}{2}(p + p_0)(v - v_0), \tag{9.5}$$

there results, in analogy to (3.23),

$$\frac{p}{p_0} = \frac{f + 1 - (v/v_0) + (2Q/p_0 v_0)}{(f + 1)\, v/v_0 - 1} = \frac{f + 1 + (f + 2)\,\bar{Q} - (v/v_0)}{(f + 1)\, v/v_0 - 1}. \tag{9.6}$$

Taking $\bar{Q}$ or Q as a parameter Eq. (9.6) determines a family of Hugoniot curves $\mathscr{H}_Q$. Points on these curves give the possible final states for reactions with reaction energies Q. In this particular case all Hugoniot curves $\mathscr{H}_Q$ have the same asymptotes in a p–v diagram. There is no intersection of $\mathscr{H}_0$ and $\mathscr{H}_e$. See Fig. 9.1.

9.2 *The points of intersection with the Rayleigh line*

In order to obtain the change of state due to a shock transition with subsequent reaction (i.e. a detonation) the intersection and tangent points of the Rayleigh line with the Hugoniot curves $\mathscr{H}_0$ and $\mathscr{H}_Q$ must be determined. Writing Eq. (8.1) in the form

$$\begin{aligned} \frac{p - p_0}{v - v_0} &= -\varrho_0^2 u_0^2 = -\gamma \frac{p_0}{v_0} M_0^2, \\ \frac{p}{p_0} - 1 &= -\gamma M_0^2 \left(\frac{v}{v_0} - 1\right) \end{aligned} \tag{9.7}$$

---

* In Ref. [38] cited above the parameter $\bar{Q}$ actually was not defined explicitly; $\bar{Q}$ rather results as the ratio of the third and first similarity parameter introduced by DAMKÖHLER.

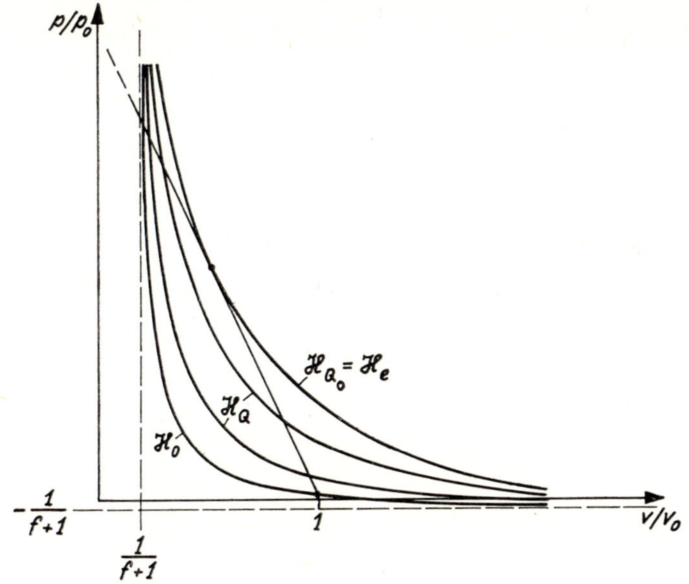

Fig. 9.1 Hugoniot curves for different reaction energies Q.

and regarding (9.4) it follows from (9.6) that

$$\left(\frac{v}{v_0}\right)_{1,2} = \frac{1}{(\gamma + 1) M_0^2} \left(\gamma M_0^2 + 1 \pm \sqrt{(M_0^2 - 1)^2 - 2(\gamma + 1) M_0^2 \bar{Q}}\right), \tag{9.8}$$

$$\left(\frac{p}{p_0}\right)_{1,2} = 1 + \left(1 - \frac{v_{1,2}}{v_0}\right) \gamma M_0^2, \tag{9.9}$$

see also [39] p. 8; [42], p. 220. It is assumed that $M_0 = u_0/a_0 > 1$ at $\xi = \xi_0$.

As expected, Eq. (9.8) yields two real intersection points of the Rayleigh line and the Hugoniot curve as far as the term under the square root is positive.

From this requirement a condition for the Damköhler parameter $\bar{Q}$ and the Mach number M_0 results

$$\bar{Q} < \frac{(M_0 - M_0^{-1})^2}{2(\gamma + 1)} \equiv \bar{Q}_{\max}(M_0, \gamma). \tag{9.10}$$

This important relation states that for an $\mathscr{R}$ process with an initial state determined by M_0 and γ only a limited amount of chemical energy up to

$\bar{Q}_{max}(M_0, \gamma)$ can be released. Equation (9.10) is also valid for $M_0 < 1$ and also for a corresponding amount of heat added from external sources*. During the heat increase, transition (1) → (2) or (0) → (2′) in Fig. 7.1, the Mach number approaches the value one. Further heat increase at $M = 1$ causes *thermal choking* and one of the premises of the $\mathcal{R}$ process, in general its steady character, is no longer satisfied. The right hand side of Eq. (8.6) tends towards infinity. This case always occurs where the Rayleigh line becomes tangent to the isentrope with $\Sigma \neq 0$. For further discussion of this phenomenon see Section 13.

Equation (9.10) may be written in a different form

$$M_0 \geq \sqrt{\frac{\gamma+1}{2}\bar{Q}+1} + \sqrt{\frac{\gamma+1}{2}\bar{Q}}\,. \tag{9.11}$$

A heat increase corresponding to $\bar{Q}$ behind the shock wave is only possible if the Mach number of the flow in front of the shock satisfies condition (9.11).

With $v/v_0 = \varrho_0 u_0 u/\varrho u u_0 = u/u_0$ instead of (9.8) we may also write

$$u = u_0 - u_0 \frac{M_0^2 - 1}{M_0^2(\gamma+1)}\left[1 \pm \sqrt{1 - \frac{2(\gamma+1)M_0^2}{(M_0^2-1)^2}\bar{Q}}\right]. \tag{9.12}$$

Plotting u as function of the parameter $\bar{Q}$ according to Eq. (9.12) a curve as shown in Fig. 9.2 results; see v. KÁRMÁN [40]. For each $\bar{Q} < \bar{Q}_{max}$ two values of u exist according to (9.12), one for the supersonic and one for the subsonic range. A transition from the supersonic to the subsonic range may occur in a stationary shock wave.

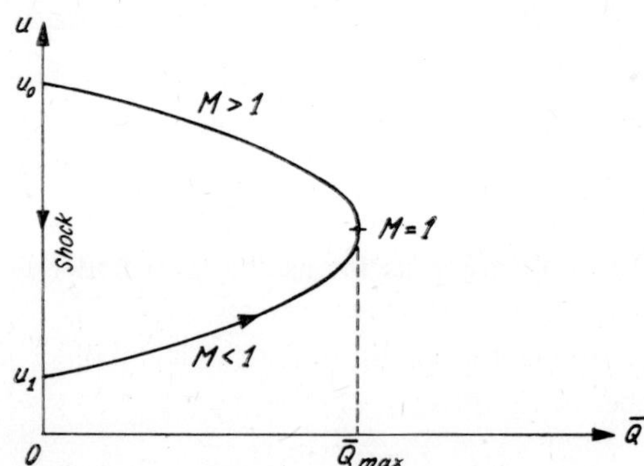

Fig. 9.2 One-dimensional flow with heat increase.

---

* In this regard it would be proper to differentiate the two mechanisms of heat increase by using the terms *heat release* and *heat supply* respectively instead of heat addition.

9.3 *Other parameters*

A somewhat different description results if for the released reaction energy instead of the Damköhler parameter a heat related to the stagnation enthalpy is used [41]:

$$\bar{\xi} = \frac{E_{ch}}{H_r} \tag{9.13}$$

with

$$H_r = H_{th} + E_{ch} + E_{kin} = \text{const.} \tag{9.14}$$

The reaction variable $\bar{\xi}$ in the case of a CJ detonation changes between the limits

$$\sqrt{\bar{\xi}_0} = \frac{M_0^2 - 1}{\gamma M_0^2 + 1} < \frac{1}{\gamma} \tag{9.15}$$

and

$$\bar{\xi} = \tilde{\bar{\xi}} = 0.$$

This parameter turns out to be more convenient for the description of the change of state during a reaction than the Damköhler parameter $\bar{Q}$.

Assuming a Chapman–Jouguet detonation we obtain for the ratios of the variables of state in the reaction zone to their values at the Chapman–Jouguet plane

$$\frac{p}{p_J} = 1 + \gamma\sqrt{\bar{\xi}}, \tag{9.16}$$

$$\frac{v}{v_J} = \frac{u}{u_J} = 1 - \sqrt{\bar{\xi}}, \tag{9.17}$$

$$M^2 = \frac{1 - \sqrt{\bar{\xi}}}{1 + \gamma\sqrt{\bar{\xi}}}. \tag{9.18}$$

Here and below CJ values are always marked by the index J.

We will call a *Chapman–Jouguet plane* the flow cross-section behind the detonation front where the flow assumes the Mach number $M = 1$, that means where the chemical equilibrium is established.

Ref. [41] investigates for such a model the stability of a Chapman–Jouguet detonation with respect to transversal disturbance waves. For the reaction velocity an Arrhenius formula is assumed. Certain instability regions are found as dependent on the length of the reaction zone; see also Section 13.

In order to obtain a simple formal distinction between the different detonation processes for the three possible cases discussed in Section 7, ADAM-

SON and MORRISON [43] introduced a dimensionless parameter for the heat increase which is defined by*

$$f = \frac{2(\gamma + 1)}{(M_0^2 - 1)^2} M_0^2 \frac{Q}{c_p T_0} = \frac{\bar{Q}}{\bar{Q}_{\max}(M_0, \gamma)} \tag{9.19}$$

or the slightly varied quantity*

$$F = 1 + \sqrt{1 - f}. \tag{9.20}$$

The parameter F is also used in Ref. [39]. F corresponds to the bracket term in formula (9.12). This results in the following classification:

$0 < f < 1;\ 1 < F < 2$	Strong Detonation,
$f > 1;\ F$ imaginary	No steady Solution,
$f = 1;\ F = 1$	CJ Detonation,
$f = 0;\ F = 2$	Shock Wave $(Q = 0)$.

For the pressure change the following relation is obtained

$$\frac{p}{p_0} = 1 + \frac{F\gamma}{\gamma + 1}(M_0^2 - 1). \tag{9.21}$$

Near the CJ point along $\mathscr{H}_e$ the pressure changes more rapidly than the upstream velocity. According to [44] p. 11

$$1 - \frac{D_{\mathrm{CJ}}^2}{D^2} \approx \left(1 - \frac{p_{\mathrm{CJ}}}{p_2}\right)^2. \tag{9.22}$$

Some improvements of the perfect-gas model were obtained in Ref. [44]. On account of the high density of the reaction products the calculation was based there on an Abel equation of state

$$p(v - b) = nR_a T$$

with a constant covolume b. See also Section 16. The cited reference contains the derivations of all basic formulas for normal and oblique shock waves with and without heat release. p_0 indeed is neglected in [44] and the results

* The quantities f and F here have of course a different meaning than those used before. A negative sign of the square root yields a weak detonation, which, however, will not be discussed furthermore.

are therefore to be considered as a good approximation only for shock pressures $p \gg p_0$—a condition frequently very accurately satisfied for detonations.

10 Chapman–Jouguet Detonations

The Chapman–Jouguet condition states that behind the reaction front a plane perpendicular to the flow direction exists where the flow velocity becomes equal to the local velocity of sound *(Chapman–Jouguet plane)*. That means

$$\left(\frac{\partial p}{\partial v}\right)_S = \frac{\Delta p}{\Delta v} = -\theta^2.$$

For the model discussed in Section 9 in the case of a Chapman–Jouguet detonation the following relations are obtained; see e.g. [45] p. 804, and [39]. Introducing the shock Mach number M_S according to (7.6) we have here $M_S = M_0$ but we will write M_S instead of M_0 since for the oblique detonation, to be discussed later, the formulas remain valid for M_S, however, not with M_0. We obtain, see also (9.10), (9.11)

$$\bar{Q} = \frac{(M_{SJ} - M_{SJ}^{-1})^2}{2(\gamma + 1)}, \quad M_{SJ} = \sqrt{\frac{\gamma + 1}{2}\bar{Q} + 1} \pm \sqrt{\frac{\gamma + 1}{2}\bar{Q}}, \tag{10.1}$$

$$M_{SJ}^2 = \frac{v_0}{\gamma p_0}\theta_J^2 = (\gamma + 1)\bar{Q} + 1 \pm \sqrt{(\gamma + 1)^2 \bar{Q}^2 + 2(\gamma + 1)\bar{Q}}, \tag{10.2}$$

$$\frac{v_J}{v_0} = 1 + \bar{Q} \mp \bar{Q}\sqrt{1 + \frac{2}{(\gamma + 1)\bar{Q}}} = 1 - \frac{1 - M_{SJ}^{-2}}{\gamma + 1}, \tag{10.3}$$

$$\frac{p_J}{p_0} = 1 + \gamma\bar{Q}\left(1 \pm \sqrt{1 + \frac{2}{(\gamma + 1)\bar{Q}}}\right) = 1 + \gamma\frac{M_{SJ}^2 - 1}{\gamma + 1}. \tag{10.4}$$

The upper sign in the above formulas applies to detonations, the lower sign to the maximal possible flame velocity in case of a combustion which would be a fictitious CJ combustion.

For large pressure ratios $p_J/p_0 \gg 1$ as in general obtained in a detonation wave, some approximate formulas can be derived which describe the general behavior of the characteristic quantities:

$$\begin{aligned} u_0 &= \sqrt{2(\gamma^2 - 1)Q} = D_J, \\ M_{SJ}^2 &= 2(\gamma + 1)\bar{Q} \gg 1, \\ \frac{D_J}{a_J} &= \frac{\varrho_J}{\varrho_0} = \frac{\gamma + 1}{\gamma}, \\ \frac{p_J}{p_0} &= 2\gamma\bar{Q} = \frac{\gamma}{\gamma + 1} M_{SJ}^2 \gg 1. \end{aligned} \tag{10.5}$$

For the limiting case $M_S, \bar{Q} \to \infty$ mentioned before, a comparison between a detonation wave (D) and a non-reactive shock wave (S) in a perfect gas with the same shock Mach number leads to the following ratios, [11] p. 318,*

$$\left.\begin{aligned} \frac{\varrho_{2D}}{\varrho_{2S}} &\approx \frac{2}{2 + f}, \\ \frac{p_{2D}}{p_{2S}} &\approx \frac{1}{2}, \\ \frac{T_{2D}}{T_{2S}} &\approx \frac{2 + f}{4}, \\ \frac{u_{2D}}{u_{2S}} &\approx \frac{1}{2}. \end{aligned}\right\} \tag{10.6}$$

11 The Structure of the Reaction Zone

Assuming that a stable Chapman–Jouguet detonation exists, the calculation of the above-mentioned characteristic quantities in general requires that the reaction energy is known, e.g. in form of the parameter $\bar{Q}$, and that the chemical equilibrium can be determined.

Whether a detonation of the substance is actually possible or not, that means the question about detonability, detonation limits, and stability can not be answered based on equilibrium thermodynamics alone. For a complete description of the detonation, the kinetics of the reaction process must also be known to some extent. The reason for this is the basic influence of

---

* Here f again stands for the number of degrees of freedom; see (2.21′).

quantities like activation energy, reaction time, or length of the reaction zone on the detonation process.

In most cases the comprehensive description of the reaction kinetics is not yet possible due to the complicated, chain-type reaction processes.

11.1 *Arrhenius' formula*

For homogeneous elementary reactions (frequently involving radicals) in many cases the constant of the reaction rate can be sufficiently accurately indicated by an *Arrhenius formula*, see e.g. [41]; [19], p. 111.

Assuming a monomolecular irreversible reaction we may write

$$k = k_0 \, e^{-E_{act}/RT}, \quad k_0 = \text{const} \tag{11.1}$$

with k as so-called reaction-rate constant and E_{act} the activation energy. Using $\bar{\xi}$ according to (9.13) with $\tilde{\bar{\xi}} = 0$ we obtain

$$r = \frac{d\bar{\xi}}{dt} = -k\bar{\xi} = -\bar{\xi}k_0 \, e^{-E_{act}/RT} \tag{11.2}$$

and for a perfect gas, see (6.1),

$$r = -\bar{\xi}k_0 \, e^{-E_{act}/pv}. \tag{11.3}$$

Near the equilibrium state it follows that, approximately, see (6.2),

$$\bar{\xi} \sim e^{-kt} \tag{11.4}$$

by integration of the differential equation (11.2).

Assuming an Arrhenius formula the variation of the state variables across the reaction zone may be calculated.

A more realistic example taken from [32] is shown in Fig. 11.1. This diagram is based on a more complicated reaction mechanism which particularly leads to an induction zone heading the actual reaction.

11.2 *Activation energy*

The existence of an activation energy is closely related to the occurrence of a reaction inhibition. This relationship may be explained simply by means of an energy level diagram, Fig. 11.2.

The only molecules participating in the reaction are those which exceed a certain energy level, namely $E_1 + E_{act}$. Since we assume that the molecules of the gas with an internal energy E_1 are subject to a Maxwell-Boltzmann

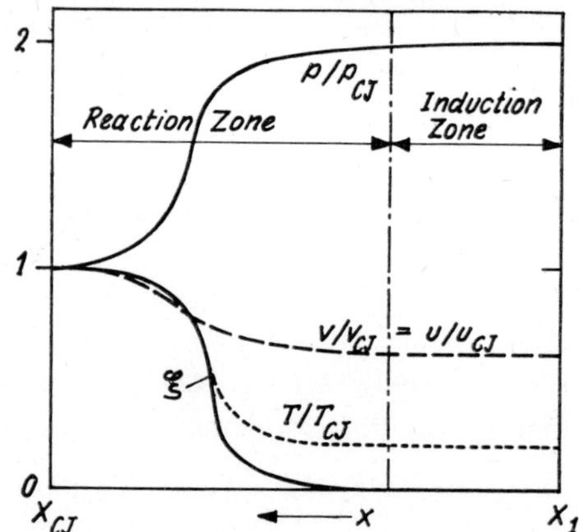

Fig. 11.1 Change of states in a detonation wave (from Ref. [32]).

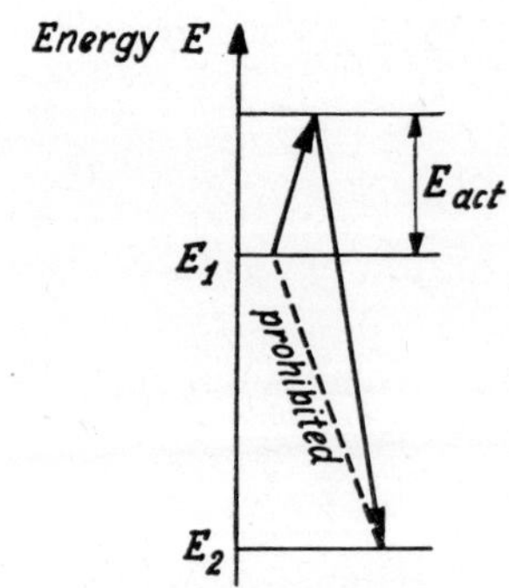

Fig. 11.2 The meaning of activation energy.

distribution the portion of the molecules having an additional (kinetic) energy E_{act} is determined by a Boltzmann factor, see (11.1). For a more detailed discussion see e.g. [47] p. 384.

11.3 *The induction period*

In the case of explosion processes the equilibrium state is in general very quickly approached. This permits the definition of a finite reaction time or reaction length. This characteristic time or length sometimes depends to a considerable extent on the occurrence of an induction period, which is not taken into account in the above formula (11.3).

During the induction period the radicals are formed, and when they exceed a certain concentration they initiate the actual chain reaction. The heat release during the induction period is in general very small. In Fig. 11.3 the influence of an induction period on the temperature increase is shown schematically for a heat explosion.

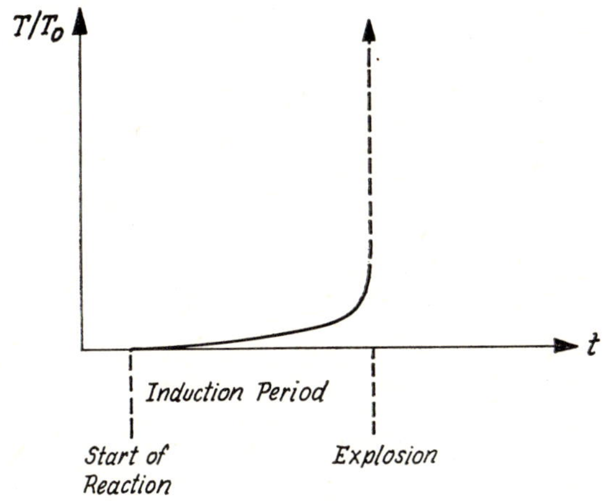

Fig. 11.3 The induction period in case of a heat explosion (from Ref. [46]).

11.4 *Homogeneous and heterogeneous reaction types*

The model induction period-Arrhenius formula describes qualitatively the reaction-zone structure with sufficient accuracy in those cases where the reactions proceed in a homogeneous fluid phase (possibly including the precipitation of a finely dispersed condensed phase of small mass). This is essentially true for detonations in gases and liquids. In solid explosives, however, the substance breaks into small grains which (as for gun powder) burn with a pressure-dependent linear velocity, inwards from the surface. The still unreacted solid substance is not in temperature equilibrium with the gas phase.

Due to the fact that the reaction first starts at sporadic *hot spots* (see Chapter IX), here too an induction period may occur. Figure 11.1 is related to such a reaction type.

We shall encounter reactions of homogeneous and heterogeneous types again but in a slightly different sense in connection with the discussion of the nonideal detonation (see Section 29.5).

CHAPTER V

The Chapman-Jouguet Condition

12 The Classical Argumentation

INVESTIGATIONS of detonating substances show that in all cases observed a selfsupporting detonation propagates with a constant detonation velocity which is only dependent on the particular properties of the detonating substance. This property represents at the same time the condition which is required for a quasi-steady treatment of the detonation process as done so far. Therefore, regarding the unavoidable transient processes during the initiation phase, this obviously means that the detonation is a stable process*.

12.1 *The Chapman–Jouguet hypothesis*

An explanation of this typical behavior of a detonation has already been given by CHAPMAN and JOUGUET in 1899 and 1905 respectively.

They assume that at a finite distance behind the detonation front, at the so-called *Chapman–Jouguet plane*, the reaction has been completed and that among the possible final states along the $\mathscr{H}_e$ curve (see Fig. 12.1), the particular state reached is the one which leads to the smallest value of the detonation velocity.

This means: For a selfsupporting detonation the respective Rayleigh line is according to CHAPMAN and JOUGUET tangent at the equilibrium Hugoniot curve. The contact point has been designated as CJ.

According to the results from Section 8 the flow velocity at CJ is equal to the local velocity of sound. Therefore, under steady conditions,

$$a_{CJ} = u_{CJ}, \tag{12.1}$$

and in the unsteady reference system

$$D_{CJ} = u'_{CJ} + a_{CJ}. \tag{12.2}$$

---

* Here we mean of course the stability against changes of the equilibrium position caused by external disturbances.

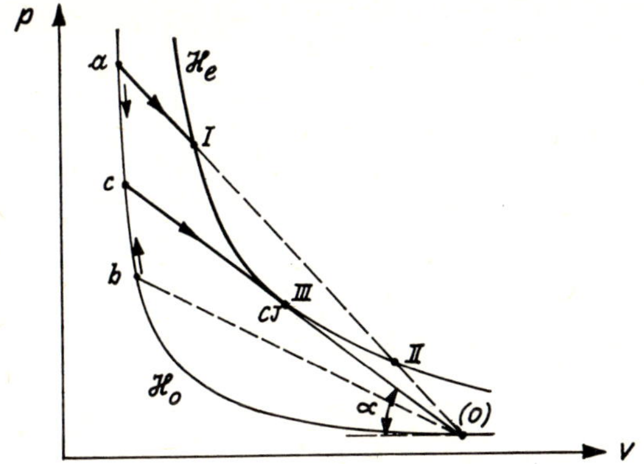

Fig. 12.1 Sketch concerning the Chapman–Jouguet hypothesis.

$$D = v_0 \sqrt{\tan \alpha}, \quad \tan \alpha = -\frac{\Delta p}{\Delta v}.$$

Relation (12.1) is usually called the *Chapman–Jouguet condition.*

Stating that the Chapman–Jouguet detonation is a stable process implies that a detonation front propagating with a higher velocity than D_{CJ} is decelerated and that vice versa for smaller velocities than D_{CJ} the wave is accelerated. In the following we shall discuss the validity of these statements.

Changes of the propagation velocity of shock waves and therefore also of detonation waves are accompanied by equivalent changes of their intensity or pressure jump, Fig. 12.1.

12.2 *Attenuation of shock waves*

The intensity of a plane shock front, e.g. defined by the pressure ratio p_2/p_0, is if considered in a stationary frame of reference determined by the upstream velocity u_0 or in a fluid at rest by the propagation velocity of the shock front. For perfect gases, e.g.

$$\frac{p_2}{p_0} = 1 + \frac{2\gamma}{\gamma + 1} \left(\frac{u_0^2}{a_0^2} - 1 \right). \tag{12.3}$$

The nonideal behavior of the matter (viscosity, thermal conduction, relaxation) has an influence only on the thickness of the shock transition layer but not on the shock intensity.

This description of the behavior of a shock wave is correct provided that the flow behind the shock front is maintained by some mechanism, e.g. a piston moving with the velocity $\hat{u}$. Such a pure shock wave is not attenuated, see Fig. 12.2.

If the pressure behind the shock front remains constant only in a limited region followed by a pressure decrease, then the total wave character is essentially unsteady. We call such a wave a pressure wave with finite amplitude,

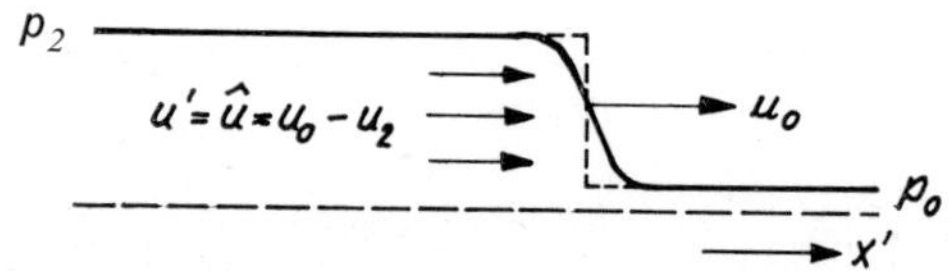

Fig. 12.2 Plane shock front followed by a constant pressure (and velocity) level.

in this particular case consisting of a shock front followed by a constant pressure level and a continuous expansion down to a pressure $p_3 = p_0$, see Fig. 12.3.

A pressure wave of this type is characterized by the fact that small pressure disturbances (sound waves) in the gas behind the shock front propagate with the velocity $u' + a$ where u' is the flow velocity of the gas and a the local velocity of sound*.

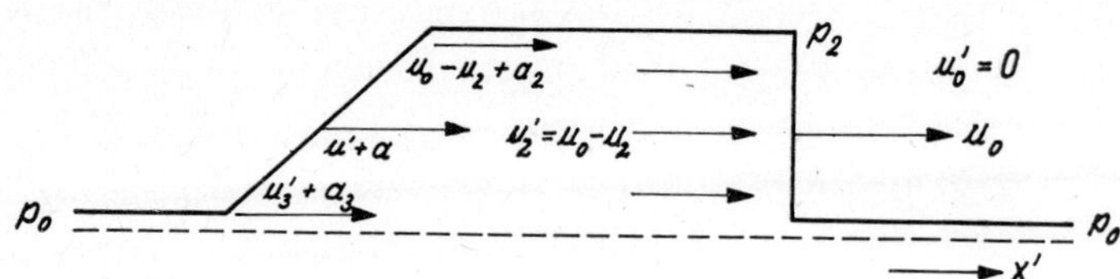

Fig. 12.3 Plane shock front with succeeding rarefaction wave (unsteady; gas at rest in front of the shock).

The relationship between the variables u and ϱ within the wave is given by

$$u - u_2 = u_2' - u' = \int_{\varrho \atop s=\text{const}}^{\varrho_2} \sqrt{\left(\frac{\partial p}{\partial \varrho}\right)_s} \frac{d\varrho}{\varrho} \quad \text{(Riemann 1860).} \tag{12.4}$$

For perfect gases (12.4) becomes with $p\varrho^{-\gamma} = p_2\varrho_2^{-\gamma} = \text{const}$

$$u = u_2 - \frac{2a_2}{\gamma - 1}\left\{1 - \left(\frac{\varrho}{\varrho_2}\right)^{(\gamma-1)/2}\right\} \tag{12.5}$$

* In a frame of reference moving with the shock front the propagation velocity would be $u - a = u_0 - (u' + a)$, see Section 1.3.

and with $a = a_2 (\varrho/\varrho_2)^{(\gamma-1)/2}$

$$u = u_2 - \frac{2}{\gamma - 1}(a_2 - a). \tag{12.6}$$

The head of the expansion wave propagates with the velocity

$$u_2' + a_2$$

and its tail with the velocity

$$u_3' + a_3.$$

For shock waves with small amplitude, with $p_3 = p_0$ also $\varrho_3 \approx \varrho_0$, and $a_3 \approx a_0$, $u_3' \approx 0$.

In each case:

$$u_3' + a_3 < u_0 < u_2' + a_2 \tag{12.7}$$

that means the constant pressure level region diminishes and the total length of the wave increases.

The intensity of the shock wave remains therefore constant until the head of the expansion wave overtakes the shock front. From this instant on a decrease of intensity is connected with the propagation of the shock front; see Fig. 12.4.

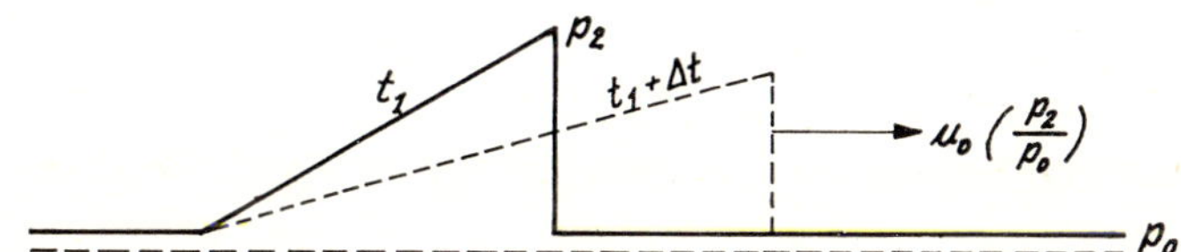

Fig. 12.4 Distortion of a "saw-tooth-type" pressure wave with shock front.

When $t \to \infty$ the pressure discontinuity decreases asymptotically towards zero if the trailing end of the wave is governed by the boundary condition $p = p_0$ as in the present case, or $u' = 0$.

This type of wave attenuation is caused only by the irreversible processes in the shock transition region and may be quantitatively calculated from the discontinuity conditions (Hugoniot equation) without explicit knowledge of the transport coefficients (which we neglect within the scope of this discussion).

Mechanical energy from the wave is also continuously lost in the first phase when the pressure value behind the shock front is still constant. This becomes obvious from the reduction of the length of the constant pressure level region.

A plane isentropic pressure wave without discontinuities (shocks) on the other hand is not attenuated on the same conditions. Amplitude and extent of the wave remain constant. However, an affine deformation of the wave pattern is observed, leading to a steepening of the compression part and a flattening of the rarefaction part; see sketch shown in Fig. 12.5.

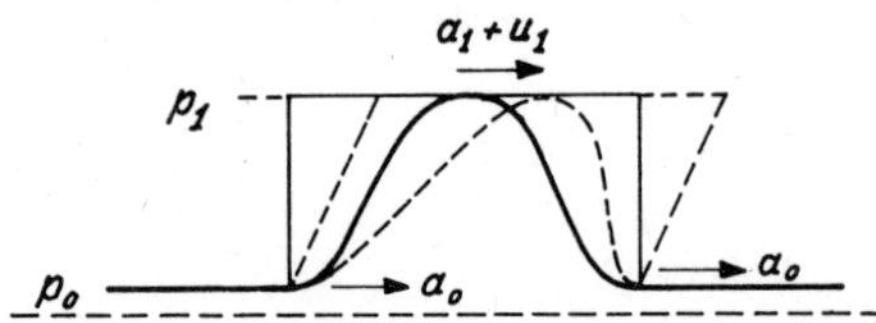

Fig. 12.5 Distortion of an adiabatic pressure wave with finite amplitude (non-linear propagation).

The deformation of the wave profile described above always leads for originally isentropic waves after some time to the generation of discontinuities and consequently to a dissipation of mechanical energy: [2], [8].

12.3 *Arguments for the stability of the Chapman–Jouguet detonation* *

We are interested now in the question of the stability of a selfsupported plane detonation front regarding its propagation velocity and the states immediately behind the wave front without taking into account the particular structure of the front itself.

For a steady detonation consisting of a shock wave followed by a reaction with changes of state along the Rayleigh line $\mathscr{R}$ starting on $\mathscr{H}_0$ and ending on $\mathscr{H}_e$ (see Fig. 12.1) the following three basic cases are conceivable (see also the discussion in 7.1):

$$\left.\begin{array}{lll} \text{I)}\ D < u_2' + a_2, & p > p_{CJ}, & \text{strong detonation,} \\ \text{II)}\ D > u_2' + a_2, & p < p_{CJ}, & \text{weak detonation,} \end{array}\right\} D > D_{CJ} \tag{12.8}$$

$$\text{III)}\ D = u_2' + a_2 = D_{CJ}, \quad p = p_{CJ}, \quad \text{CJ detonation.} \tag{12.9}$$

Here D is the propagation velocity of the detonation wave, D^2 being proportional to the negative slope of $\mathscr{R}$ (Fig. 12.1), and $u_2' + a_2$ is the propagation velocity of isentropic pressure waves in the same direction within the reaction products behind the reaction zone.

* This discussion corresponds approximately to the now already classical physico-phenomenological argumentation for the stability of the Chapman–Jouguet detonation found in the literature, [28] Section 8, [29], [51] [63].

The case $D < D_{CJ}$ is not possible as a steady process. See case b) in Fig. 12.1; $\mathscr{H}_e$ is not attained.

Resulting from fluid-mechanic and thermodynamic reasons, only case III is stable based on the following argumentation:

I. The strong detonation with subsonic flow behind the reaction zone is subject to the aforementioned features of shock wave attenuation. The wave intensity is decreased by succeeding expansion waves penetrating the reaction zone provided that no piston motion maintains p_2 and u'_2.

II. Regarding the case of a weak detonation it is assumed that during the detonation process all intermediate states normally fall into the p–v region between $\mathscr{H}_0$ and $\mathscr{H}_e$ (Fig. 13.2). Consequently the p–v region beyond $\mathscr{H}_e$ is excluded and the weak detonation branch of $\mathscr{H}_e$ not attainable from above along $\mathscr{R}$ (e.g. from point a along line II in Fig. 12.1). A weak detonation is normally not possible.

III. The CJ detonation is left as the only possible stable case and must consequently be identical to the empirically confirmed stable, steady, self-supported detonation.

We make a few additional remarks regarding the three cases listed above.

Regarding case I: A typical example for this case is the initiation at the closed end of a tube containing an explosive or a reactive gas mixture. The detonation may be initiated by an externally generated shock wave. In consequence of the mass balance for the detonation process sketched in Fig. 12.6, a rarefaction follows the compression. This example will be treated in detail in Section 23 (case $\nu = 1$).

Regarding case II: The weak detonation branch of the $\mathscr{H}_e$-curve may be attained, a) without shock transition by means of a supersonic combustion or a condensation, compare Chapter III; b) in the so-called *pathological case* where the region of intermediate states reaches beyond $\mathscr{H}_e$; see the following discussion in Sections 13 and 14.

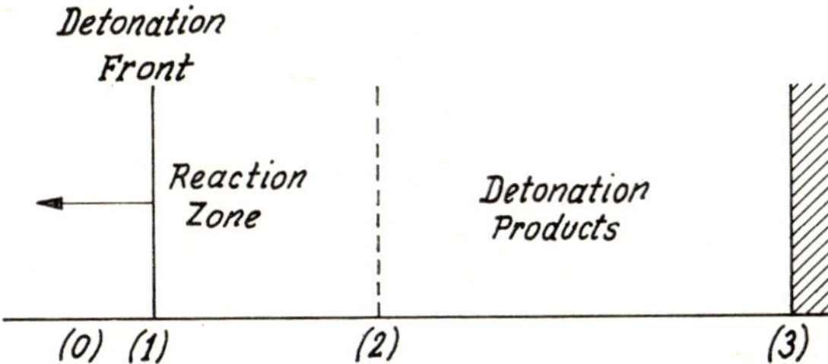

Fig. 12.6 Propagation of a plane detonation wave into a semi-infinite space bounded by a rigid wall (3); $u_3 = 0$ is consequently a boundary condition.

Regarding case III: The objections for the cases I and II do not exist in this case. This is all we can state presently.

Various arguments listed elsewhere which are no longer tenable are not discussed (like e.g. entropy arguments or the supposed instability of a wave with supersonic downstream flow).

13 Criticism of the CJ Hypothesis, Limitations

In a more thorough discussion of the detonation process the following points have to be considered:

a) In our preceding argumentation regarding the CJ hypothesis we had to assume an expansion wave following the CJ plane providing the transition from the detonation wave to the final state at the boundary of the detonating substance.

The CJ plane, however, does not exist in the ideal case, or we may say is shifted to infinity for asymptotic equilibrium approach. This is obviously a contradiction. But as a limiting case and for hypothetical experiments the Chapman–Jouguet hypothesis remains correct.

After running a sufficiently long time the expansion behind the detonation wave, initiated at the plane rigid boundary and propagating into the explosive semispace becomes arbitrarily flat. The steady ideal CJ-zone structure is then asymptotically obtained for $t \to \infty$. Also, see the pseudostationary solution discussed in Section 23.

b) The stability of the CJ detonation requires, as mentioned before, that a too slowly propagating shock front ($D < D_{CJ}$) is accelerated by the wave mechanism itself. Is this always true?

We know that a shock with a following steady reaction zone propagating with $D < D_{CJ}$ is impossible. We may therefore ask for the exact point where the theoretical calculated construction of this zone is no longer feasible, and how the mechanism of the self-acceleration is connected with this feature.

Following the discussion in Section 8 the reaction-zone structure results from the integration of the differential Eqs. (4.3), (8.6), (8.7) starting from the shock. We consider particularly (8.6) or

$$\frac{1}{v}\frac{dv}{dt} = \frac{\Sigma}{1 - M_f^2}. \tag{13.1}$$

We assume at first the "*normal*" *case*, namely that during the entire reaction

$$\Sigma \equiv \sum_j \sigma_j r_j > 0 \tag{13.2}$$

(expansive reaction). It can be realized that the purely mathematical zone construction by integration is possible up to the equilibrium state if during the entire reaction

$$M_f < 1 \tag{13.3}$$

and only on this condition; Eq. (13.3) is satisfied immediately behind the shock front. Otherwise the integration ends at a point $t = t^*$ where

$$M_f = 1, \quad \Sigma > 0. \tag{13.4}$$

Here (13.1) leads to a singularity, allowing no continuation for $t > t^*$ but joining two branches with $t < t^*$.

This mathematical behavior corresponds to the well known effect of thermal choking in a diabatic pipe flow, see Section 9.2. The resulting unsteady pressure waves are responsible for the shock wave acceleration; see Fig. 12.1, case b), and Ref. [48].

Unanswered remains the question why the acceleration caused by thermal choking according to (13.4) should lead to a stable limiting configuration which now is no longer defined by M_f but rather by M_e, namely the CJ_e detonation. We will return to this problem later. The classical theory does not distinguish between M_f and M_e. Compare in this respect Figs. 13.1 and 13.2.

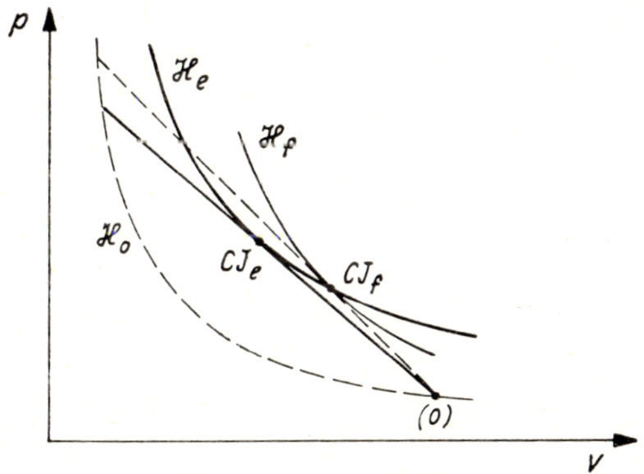

Fig. 13.1 Hugoniot curves for the case $a_f > a_e$.

c) The objections already raised by von Neumann [50] known as *von Neumann's pathological case* are formulated here in a slightly different way. If at the beginning of the reaction (13.2) and (13.3) are valid, later on, however, Σ and $1 - M_f^2$ pass through zero simultaneously and are negative from thereon, dv/dt remains continuous and >0 according to (13.1). The equilibrium is approached with $M_f > 1$ and consequently all the more with $M_e > 1$ (case of weak detonation). This possibility will also be discussed later.

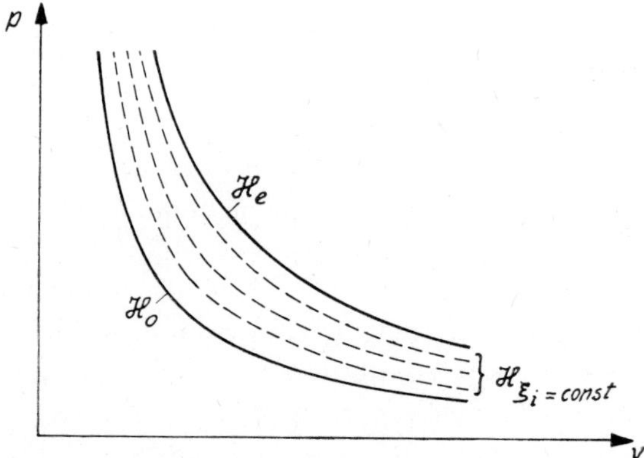

Fig. 13.2 Hugoniot curves for the case $a_f \equiv a_e$.

d) Instabilities of a plane detonation wave due to transversal or longitudinal pressure disturbances have been found and investigated in many cases [31], [49], [53], [61]. As far as those instabilities are present, the assumption of a one-dimensional steady and laminar zone structure does not hold true and the CJ condition is violated. Instead, wave patterns of two or three dimensions appear, with frequently occurring spatial and temporal periodicity (spinning detonation). These phenomena were described and investigated in many papers and some monographs, e.g. Ref. [56], [52]. Only on an average the process is one-dimensional and steady, that means only after taking the mean values by integrating over y, z and t the flow pattern will depend solely on x and the velocities be in the x-direction.

After a calming zone a constant average final state with $\bar{p}$, $\bar{v}$, $\bar{u}$ will be established and we may ask whether this state together with the average detonation velocity $\bar{D}$ satisfies the CJ condition. This is *not* to be expected since the reasoning leading to the CJ condition is not valid for the averaged states. The averaged states, especially, are not restricted to the *Rayleigh line* which is determined by the initial and final states. Therefore the argument II on page 79 by which a weak detonation was excluded is here no longer applicable.

e) In [62] the CJ condition is contested for a special case based on experimental results. (According to [67] this discrepancy between theory and experiment is due to an incorrect evaluation in the experimental determination of the detonation pressure.) This supports the theoretical objections made above, however, it does not lead to a clarification. In [31] the deviation is suspected to be due to longitudinal oscillations.

[62] uses basically the inverse method according to [65] and [66]; see Section 19.

14 Present Status

The outlines in 13 have left open the discussion of the objections listed under points b) and c) which possibly might lead to a modification of the CJ condition.

Regarding point b) we add the following idea: while the discussion for $D > D_{\mathrm{CJ}_e}$ in Section 12 for the normal case necessarily leads to the CJ_e condition, only M_f appears in the range $D < D_{\mathrm{CJ}_e}$. This explains the fact that some authors [25], [57], at least for some time, have considered a CJ_f condition with $M_f = 1$ at the final state as conclusive. This would mean that the stable end point with $D = D_{\mathrm{CJ}_f} > D_{\mathrm{CJ}_e}$ is found on the weak detonation branch with $M_e > 1$, see Fig. 13.1. The corresponding $\mathscr{R}$ line would be tangent to the frozen Hugoniot curve $\mathscr{H}_f$ belonging to this state.

Later on [58], [59] this assumption was again abandoned and the CJ_e condition considered as valid exclusively. This is (with exemption of [62]) also in agreement with the experimental results proved by measurements of the

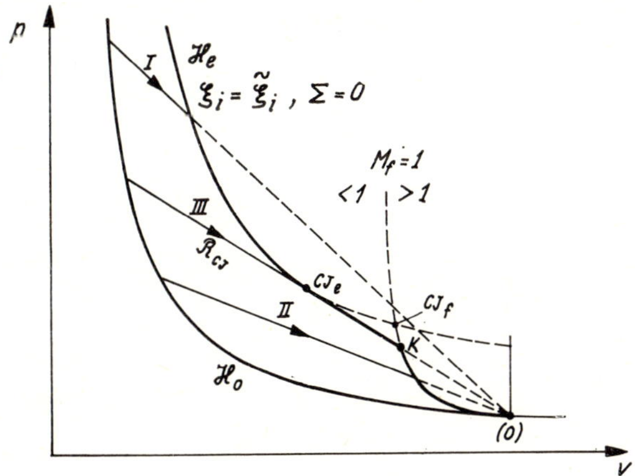

Fig. 14.1 Reaction process with $\Sigma \geqq 0$; normal development of the detonation with CJ_e as a stable final state (compare Fig. 14.5).

detonation velocity [55], p. 107, the pressure [56] and the density [54] behind the detonation front in detonating gas mixtures. The situation, however, was still unsatisfactory; this becomes particularly noticeable in the discussion of Ref. [60]. Ref. [64] reports the calculation of the CJ_e detonation for a number of gas mixtures including the values of M_f. In all cases $0.94 < M_f < 0.98$ was found.

We come to the conclusion that, in the normal case, due to the reasons listed in Section 12, only the CJ_e detonation can be stable.

Neighboring "paths" with $D < D_{CJ_e}$ (case II in Fig. 14.1) end with thermal choking at the sonic limit $M_f = 1$. The attainable region of the p–v plane is bounded by a part of $\mathscr{H}_e$ above CJ_e and of $M_f = 1$ below K connected by a piece of the $\mathscr{R}$ line (from CJ_e to K). CJ_f cannot be attained.

A model for this behavior can be constructed easily if, in the relaxation model discussed in Section 6, the assumption $\xi_0 = \tilde{\xi}_0$, which means equilibrium conditions at the initial state, is replaced by $\xi_0 > \tilde{\xi}_0$. It was there $\gamma_f = \text{const}$, $\gamma_e = \text{const} < \gamma_f$. The sonic limits $M_f = 1$ and $M_e = 1$ are hyperbolas in this case, (7.9).

The apparent jump from K to CJ_e when D approaches D_{CJ} from below is eliminated if the time is taken as a coordinate, see Fig. 14.2. The limiting position of curve II when $D \to D_{CJ}$ then coincides with curve III plus the segment $CJ_e - K$, these latter being physically meaningless on account of $t = \infty$. While the points on $M_f = 1$ below K (Fig. 14.1) are reached in finite

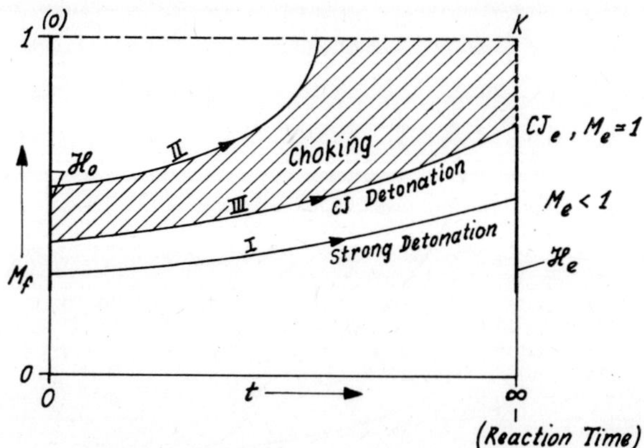

Fig. 14.2 Representation of the solution curves of Eq. (13.1) in the M_f-t plane for the three cases sketched in Fig. 14.1: I: strong detonation; subsonic downstream flow; II: acceleration of the detonation through thermal choking; III: Chapman–Jouguet detonation, sonic downstream flow with velocity a_e.

time, $\mathscr{R}_{\mathrm{CJ}}$ has, owing to (13.1), a "stagnation point" (CJ_e) which is approached only asymptotically for $t \to \infty$. Everywhere on $\mathscr{H}_e$ we have of course $\Sigma = 0$. On $\mathscr{R}_{\mathrm{CJ}_e}$ beyond CJ_e there is again $dv/dt > 0$. The reaction here only apparently leads away from the equilibrium. The equilibrium approached by the reaction is shifted continuously.

In Fig. 14.2 the second basic coordinate is M_f so that the afore-mentioned attainable region is bounded here by the two sides of a right angle at K. The CJ_e detonation is, according to this property, absolutely stable with respect to $D \gtrless D_{\mathrm{CJ}e}$.

The one-to-one correspondence between the v–p and the t–M_f plane supposed here is, however, not guaranteed and the proof is therefore not completely cogent.

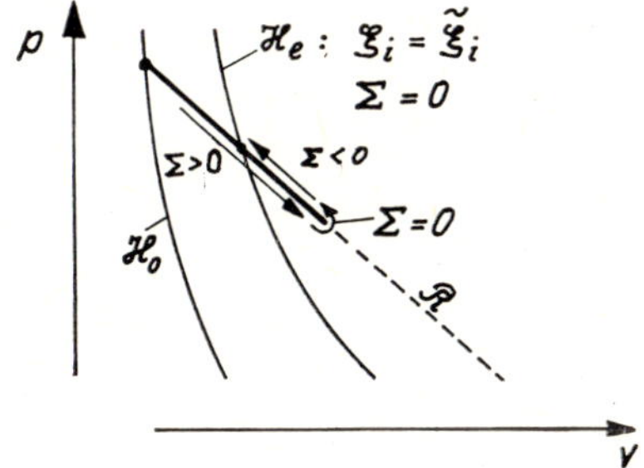

Fig. 14.3 Reaction progress along the Rayleigh line for a strong detonation with a change in sign for Σ.

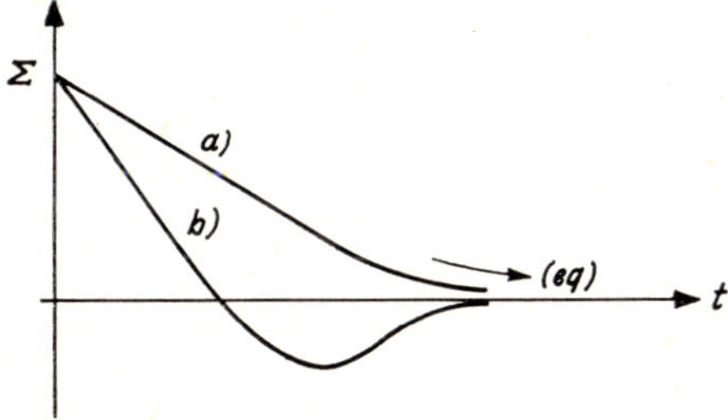

Fig. 14.4 Development of the quantity Σ during the reaction, a) normal case, b) pathological case.

Regarding c) we add this remark: suppose initially $\Sigma > 0$ and subsequently $\Sigma < 0$ and that a stable steady detonation with $D = D_{\mathrm{st}}$ exists. Certainly then the forced detonation with $D > D_{\mathrm{st}}$ driven by a propelling piston and consequently with $M_f < 1$ in the entire reaction zone also exists. According to (13.1) the state point starting from $\mathscr{H}_0$ is forced to move across

$\mathscr{H}_e$ and to return at $\Sigma = 0$ ending finally on $\mathscr{H}_e$. See Figs. 14.3, 14.5 case I. $\Sigma = 0$ is the boundary of the attainable region in the v–p plane. The region between $\mathscr{H}_e$ and $\Sigma = 0$ is thereby covered twice, one time each by states $\Sigma > 0$ and $\Sigma < 0$.

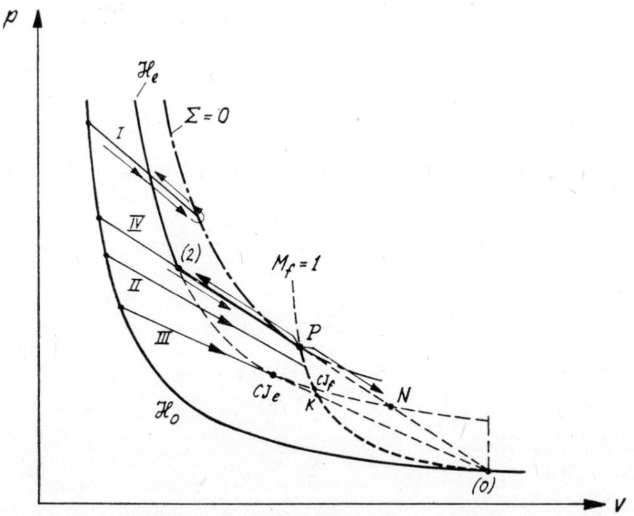

Fig. 14.5 Behavior of the detonation for a change in sign of the quantity Σ during the reaction process (compare Fig. 14.1). III: A final state on $\mathscr{H}_e$ with sonic downstream flow cannot be obtained (CJ condition violated). IV: The state N with supersonic downstream flow is a possible stable final state (v. Neumann's pathological case).

Now assume we that the steady strong detonation becomes gradually weaker until the self-supported stable detonation with $D = D_{st}$ is obtained. This wave cannot be influenced by succeeding rarefaction waves, that means there exists at least one point where $M_f = 1$.

The variation of M_f along $\mathscr{R}$ is given by

$$\frac{dM_f^2}{dt} = M_f^2 (1 + \Gamma_f M_f^2) \frac{\Sigma}{1 - M_f^2} - M_f^2 \frac{d \ln \Gamma_f}{dt} \tag{14.1}$$

with

$$a_f^2 = \left(\frac{\partial p}{\partial \varrho}\right)_{S,\xi_i}; \quad \Gamma_f = \frac{a_f^2}{pv}; \quad M_f = \frac{u}{a_f}.$$

The last term in (14.1) vanishes in the case of a perfect gas, where $\Gamma_f \equiv \gamma = \text{const}$. This term is otherwise continuous and of little influence since in general Γ_f will vary slowly. M_f therefore, due to (14.1) approaches the value 1 or moves away from this value depending on whether $\Sigma > 0$ (heat release) or $\Sigma < 0$ (heat consumption).

A "pathological" Σ variation according to Fig. 14.4 acts like a "thermal nozzle" with the throat at $\Sigma = 0$, $d\Sigma/dt < 0$. The entropy, however, increases monotonically during the reaction process. For the pathological case there is no maximum for S at $M_f = 1$; compare Sec. 8.2.

The differential equation (14.1) determines a vector field with a saddle point P at $\Sigma = 1 - M_f = 0$ (Fig. 14.6). The separatrix IV through P separates the subsonic paths (I, strong detonation, self-retarding) from the choking paths (II, unsteady, self-accelerating) and is therefore stable in the same sense as the CJ detonation in the normal case.

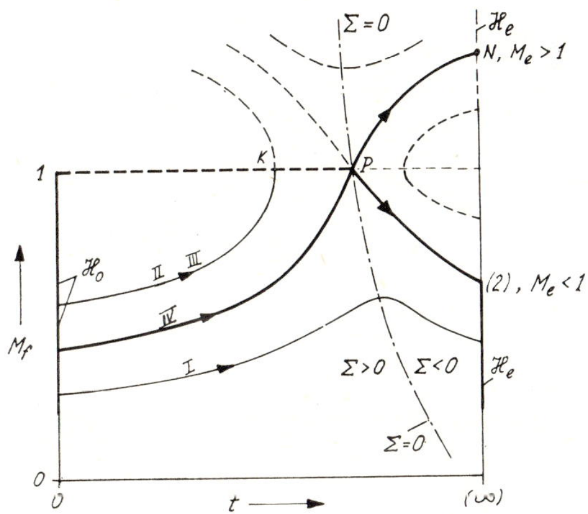

Fig. 14.6 Shape of the solution curves in the $M_f - t$ plane for the cases I, II, IV sketched n Fig. 14.5.

Beyond the saddle point P there are, similarly as for the flow through a Laval nozzle, two steady continuations possible: with $M_f > 1$, $dv/dt > 0$ or $M_f < 1$, $dv/dt < 0$. In our case with low pressure downstream the first possible continuation will be always established. The pathological $\mathscr{R}$ line IV, Fig. 14.5, runs tangentially to $\Sigma = 0$ beyond P up to the end point N on $\mathscr{R}_e$ and represents a stable weak detonation.

The straight line III $\equiv R_{CJe}$ is a choking curve which passes through the CJ_e point in the v–p plane with $\Sigma > 0$. CJ_e cannot be attained as an equilibrium point.

Whether the pathological case can be realized experimentally is not known*. Why this case could not be constructed theoretically with the two-reaction model in [59], [60] is not quite clear.

From a physico-chemical point of view certainly the reverse case is more likely: Initially $\Sigma < 0$ (cracking of chemical bonds, energy consuming) and subsequently $\Sigma > 0$ (generation of stable end products). The CJ condition would not be violated hereby as can be seen easily.

---

* According to Davis and Fickett [105] a pathological detonation seems to have been experimentally established by Venable (to be published).

CHAPTER VI

Real Explosives, Equation of State of the Reaction Products

15 Critical Conditions (Detonability)

ANY SUBSTANCE which after proper initiation is capable of carrying a self-supported detonation may be called an *explosive**. A necessary condition for the detonability of a substance is apparently that 1) the substance is in a metastable state with frozen reaction. If this is the case the equilibrium Hugoniot curve $\mathscr{H}_e$ in the *p-v* diagram does not pass through the initial state point. With the given thermodynamic conditions the curves $\mathscr{H}_0$, $\mathscr{H}_e$, and the tangent $\mathscr{R}_J$ can be drawn and the spike state (1), see Fig. 7.1, and the CJ state including the ideal detonation velocity D_J can be determined.

A further necessary condition for the detonability is then that 2) a shock wave propagating in the substance with the velocity D_J initiates the reaction. We may also say that in the spike state the circumstances hampering the reaction are removed. Both conditions together are sufficient for the explosive character of the substance. For a very lean hydrogen-air mixture, e.g., the first condition would be satisfied, not, however, the second.

Actually condition 2 is a statement regarding the length of the reaction zone. This will be taken up again in the discussion of the nonideal detonation.

16 The State Behind the Detonation Front

The common explosives with a gaseous, liquid, or solid initial state are composed of the elements C, H, O, N. The reaction products at the CJ state contain mainly the chemical species H_2O, H_2, CO_2, CO, N_2, C (as soot) and to a less extent H, O_2, OH, NH_3, NO, CH_4 and many others.

* The term explosive quite often comprises also solid propellants or gun powder. A condensed (liquid or solid) substance developed for detonation is then called a high explosive.

These components which, except for the solid carbon, are in the gaseous state, behave for normal pressures and temperatures almost like perfect gases. Also the condensation of water vapor becomes in general not effective (compare the discussion below). Under the extreme conditions, however, found behind a detonation front the real gas effects are dominating to such a degree that the perfect* gas approximation becomes inapplicable.

The order of magnitude of the values of the state variables behind the detonation front in a gaseous and solid explosive becomes apparent from Table 16.1.

Table 16.1 Comparison of the characteristic quantities of two real explosives with a gaseous [64] and solid [70] initial state respectively

	Oxyhydrogen	Composition B*
p_0	1 bar	1 bar
ϱ_0	0.49 g/dm^3	1.71 g/cm^3
D_J	2.84 km/s	7.99 km/s
p_J	19.1 bar	290 kbar (!)
u'_J	1.3 km/s	2.12 km/s
a_J	1.54 km/s	5.87 km/s
p_J/ϱ_0	1.84	1.36
$\Gamma_{J,e}$	1.128	2.77
E_0	13.8 kJ/g	5.55 kJ/g
E_J	14.7 kJ/g	7.79 kJ/g
$\bar{\Gamma}_J$	1.144	2.60
T_J	3610 K	~3000 K

The six quantities p_0, ϱ_0, D_J, p_J, E_0, T_J are formally independent from each other. The remaining six quantities can be expressed in terms of the first six by means of the three known conservation theorems, the CJ condition $D_J = a_J + u'_J$, and two additional relations†:

$$\Gamma_e = \left(\frac{\partial H}{\partial E}\right)_{S,\,\xi_i=\tilde{\xi}_i} = \frac{a_e^2}{pv} = \frac{\varrho}{p}\left(\frac{\partial p}{\partial \varrho}\right)_{S,\,\xi_i=\tilde{\xi}_i}, \tag{16.1}$$

$$\bar{\Gamma} = \frac{H}{E} = 1 + \frac{2}{f}. \tag{16.2}$$

* Grade A Composition B is a typical high explosive used for military purposes which is composed of RDX/TNT/Wax with a weight ratio 64/35/1; see [70].

† Compare (2.27); in general $\bar{\Gamma} \approx \Gamma \neq \gamma \equiv c_p/c_v$. Particularly for the condensed phase the values of γ and Γ differ considerably.

With

$$E = \frac{pv}{\bar{\Gamma} - 1} = \frac{1}{2} fpv \tag{16.3}$$

a number f, in general not integer, of quasi-degrees of freedom is defined involving particularly also the chemical energy form; compare (2.24), (2.24′). The adiabatic exponent Γ_e and the compression ratio have a close relationship. Regarding this fact compare Section 18, particularly relations (18.9), (18.11).

The quantitative difference between the values for the gaseous and solid explosive is particularly striking for the densities and pressures. Likewise the values $\bar{\Gamma}$ or $f = 2/(\bar{\Gamma} - 1)$ are different in a characteristic manner. The chemically very reactive gas, namely, has a high number of quasi degrees of freedom, while for the highly compressed reaction products of the solid explosive the mobility of the molecules is very restricted leaving correspondingly "little freedom" for any motion.

For completeness also the temperatures are listed in Table 16.1. They, however, do not appear in our relations and their CJ values are fairly uncertain for the existing high densities in case of condensed explosives.

17 **Detonation Energies**

The specific internal energy and enthalpy of a substance being in an equilibrium or nonequilibrium state is only defined up to an additive constant. This constant is determined by the definition of a certain reference state, e.g. assuming $E = 0$ for this state. Therefore, only energy or enthalpy *differences* between two states of a substance have a unique meaning.

As a reference state frequently the *normal state* 1 atmosphere and 0°C is chosen, or the ideal gas state at $T = 0$ where consequently $p = \varrho = 0$; this is termed the *zero state*. In addition the chemical state has to be described which can be done in different ways.

The enthalpy of a substance in the reference state with respect to the reference states of its elementary components is called *formation enthalpy* or *heat of formation*. With reference to the completely burned substance, that means for chemical equilibrium with oxygen in excess, we obtain the negative *heat of combustion**. Both values are well defined and can be accurately determined by calorimetric methods. The sum of the heat of formation and

* Compare [74]. Our definitions are partly different in sign from the definitions of this reference.

the combustion heat is equal to the combustion heat of all elementary components.

If reference is made to the chemical equilibrium of the decomposition products the term *heat of explosion** is used. This value is being specified in general as energy of the explosive; it is also understood in the listings of Table 16.1. The definition assumes an idealized combustion process which attains a standard equilibrium state. This energy value is not necessarily the actual detonation energy.

It is hardly necessary to distinguish between energy and enthalpy since the values of pv for the initial and final state are small.

The heat of explosion is not so well defined as the combustion heat. It depends to some extent on the experimental conditions.

As *effective detonation energy* we should, however, define only this part W of the energy released in the explosion process which can be exploited as mechanical work. The thermal and chemical rest energy remaining with the reaction products is to be regarded as lost in this respect. In order to define W we assume a lossless (isentropic) adiabatic expansion starting from the CJ-point and ending at p_0. Consequently we have for the CJ state

$$E_J = -\int_{\substack{0 \\ (S=S_J)}}^{p_J} p\, dv, \quad W_J = -\int_{\substack{p=p_0 \\ (S=S_J)}}^{p_J} (p - p_0)\, dv, \tag{17.1}$$

that means, interpreted in areas (Fig. 17.1),

$$E_J = A_1 + A_2 + A_3 + A_4, \quad W_J = A_1 + A_2.$$

From the Hugoniot equation (3.11) we have

$$E_J - E_0 = A_1 + A_3,$$

correspondingly

$$W_J - W_0 = A_1$$

whence for the initial state

$$W_0 = A_2, \quad E_0 = A_2 + A_4.$$

For condensed explosives, of course, the pressure p_0 and the areas A_3, A_4 are always negligible, yielding

$$W_J = E_J, \quad W_0 = E_0.$$

---

* Compare [74]. Our definitions are partly different in sign from the definitions of this reference.

Even the definitions (17.1) are not unequivocal since the isentropic expansion may happen either at shifting equilibrium (*eq*) or with constant composition (*fr*). Actually the adiabatic expansion will proceed at first almost at equilibrium-isentropic conditions and later at frozen-isentropic conditions. Between these two processes a (perhaps very small) non-isentropic freezing

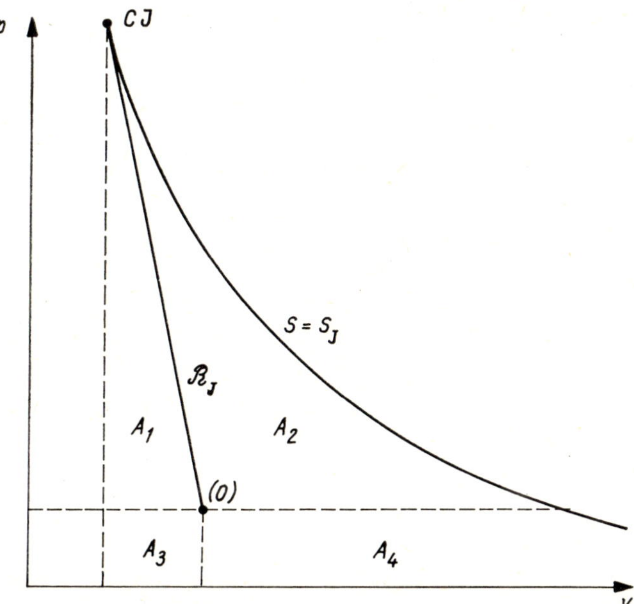

Fig. 17.1 Detonation energy and mechanical work.

zone exists involving chemical relaxations. In order to simplify we may introduce a certain *freezing temperature* (sudden freezing, constant entropy). The shape of the adiabat and the value of the mechanical work are very difficult to determine experimentally. Strictly speaking they are dependent on the expansion velocity.

For detonations of common explosives in the atmosphere occurring often with an oxygen deficit a considerable additional amount of energy is released by a succeeding combustion with the oxygen of the air [75]. It is, however, not known if this energy is released fast enough to contribute to an increased effect of the air shock.

After freezing of all chemical reactions during an adiabatic expansion in the case of phase equilibrium the gaseous reaction products completely condense until $p = T = 0$. The condensed state $p = T = 0$ has a lower energy than the gaseous state ($v = \infty$) under equal conditions. In Fig. 17.2 according to [69] this behavior is sketched schematically for a Van der Waals gas.

Due to undercooling and possibly occurring condensation shocks the actual expansion will not follow the phase-equilibrium isentrope. Moreover in explosion processes very often the condensation limit is never attained. The multiphase equilibrium isentrope for a multi-component system is difficult to determine and gives with respect to the very high detonation energy only an insignificant correction. The energy discussion of strongly expanding reaction products is therefore in general based on a fictitious ideal gas behavior neglecting the condensation and freezing of the components, even for water vapor.*

Condensation and the other phase transitions can be described also of course by the same formalism using $\xi_j, r_j, q_j, \sigma_j$ as well as the ideas of equilibrium and relaxation etc., like for the actual chemical reactions.

The terms *thermal energy* and *chemical energy* used above remain to be discussed in some more detail. If $H(p, v, \xi_j)$ is the enthalpy function of a certain substance and $E_{ch}(\xi_j)$ the value of H at the zero state, e.g. ideal gas at $T = p = 0$ with solid carbon, then $E_{ch}(\xi_j)$ is the *zero-point energy* and also the *zero-point enthalpy* of the substance. We write

$$\begin{aligned} H(p, v, \xi_j) &= H_{th}(p, v, \xi_j) + E_{ch}(\xi_j) \\ &= E_{th}(p, v, \xi_j) + pv + E_{ch}(\xi_j), \end{aligned} \tag{17.2}$$

defining in this way E_{th} and H_{th} as thermal energy and thermal enthalpy respectively, E_{ch} as chemical energy which is equal to the chemical enthalpy since $pv = 0$. At the zero state we have therefore $E_{th} = H_{th} = 0$, and for the condensed equilibrium state at $p = T = 0$, $E_{th} = H_{th} < 0$, and in the gaseous state at low temperatures $E_{th} \approx (1/\gamma)\, H_{th} \approx pv/(\gamma - 1)$. E_{ch} is a linear function of the form

$$E_{ch}(\xi_j) = \sum_j q_0^j \xi_j + \text{const}; \tag{17.3}$$

q_0^j being the j-th reaction energy at the zero state;

$$q_{pv}^j = q_{pT}^j = q_{Tv}^j = Q_{pv}^j = Q_{pT}^j = Q_{Tv}^j,$$

compare (5.12), (5.13), (5.18).

While the relations (17.2) and (17.3) are valid in general, the arrangement

$$E(p, v, \xi_j) = E_{th}(p, v) + E_{ch}(\xi_j) \tag{17.4}$$

* In [74] indeed the liquid phase of H_2O is taken into account.

represents already a considerable restricting and simplifying assumption. It means that all reaction energies

$$q^j = \frac{\partial E}{\partial \xi_j} = \frac{\partial E_{\text{ch}}}{\partial \xi_j}$$

are independent of p and v. The assumption (17.4) has been made e.g. for the *Lighthill gas** and also in Sec. 9.

A far more special assumption than (17.4) was made by (6.4) when we discussed a relaxation model.

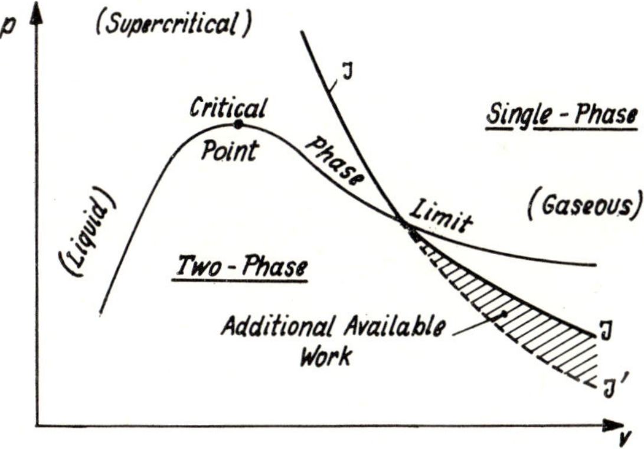

Fig. 17.2 Adiabatic expansion of the reaction products with phase transition. $\mathscr{J}$ isentrope for phase equilibrium, $\mathscr{J}'$ isentrope for undercooled gas.

18 Formulas

In the following we list some formulas for the description of detonation processes which are valid under very general conditions supplementing and generalizing the formulas for ideal gases derived in Chapter IV.

It is practical to apply here some other definitions which differ from those used in Chapter IV since some of the quantities used before, as e.g. a_0, have no essential relation to the detonation process. Thereby also the quantities $M_0, \gamma_0, H_{0\text{th}}$ are eliminated.

Regarding the initial state only the quantities v_0, p_0, and H_0 (or E_0) are here of interest†. Eventually also p_0 is omitted in addition (see below).

* A model for an ideally dissociating gas proposed by LIGHTHILL (1957, 1960).

† This is different if we ask also for the length of the reaction zone.

With $\Gamma = \Gamma_{e,J}$, $a = a_{e,J}$ according to (16.1) for equilibrium isentropic changes of state near the CJ point

$$E \approx \frac{pv}{\Gamma - 1} + \text{const.} \tag{18.1}$$

We define the constant by

$$\left(E - \frac{pv}{\Gamma - 1}\right)_{\text{CJ}} = \left(H - \frac{\Gamma pv}{\Gamma - 1}\right)_{\text{CJ}} = E^{0}. \tag{18.2}$$

E° is an integration constant and may formally be explained as zero-point energy or chemical energy. According to experience

$$E - \frac{pv}{\Gamma - 1} \equiv H - \frac{\Gamma pv}{\Gamma - 1} \approx E^{0} \tag{18.3}$$

is a good approximation along a larger part of the CJ isentrope. The reference level for E, E°, E_0 is hereby arbitrary, however, of course, identical for all three. Such a definition is reasonable because a reference state can be prescribed easily while, however, the actual zero state (obtained by an isentropic expansion from CJ to $p = 0$) is difficult to determine.

Since the isentrope and the Hugoniot curve have a contact point at CJ, (Fig. 7.5) we apply (18.3) also along $\mathscr{H}_e$ in the vicinity of CJ. Using an equilibrium state equation

$$\begin{aligned} H^{e}(p, v) &= H^{e}_{\text{th}}(p, v) + E^{0}, \\ H^{e}_{\text{th}}(p, v) &= \frac{\Gamma pv}{\Gamma - 1} \end{aligned} \tag{18.3'}$$

(compare (4.15), (9.2)) along $\mathscr{H}_e$ and even for p_0, v_0 we may define a reaction enthalpy

$$\begin{aligned} Q &= Q_{pv}(p_0, v_0) = H_0 - H^{e}(p_0, v_0), \\ Q &= H_0 - E^{0} - \frac{\Gamma p_0 v_0}{\Gamma - 1} \end{aligned} \tag{18.4}$$

and a generalized Damköhler parameter (see (3.25), (9.4))

$$\bar{Q} = \frac{Q}{H^{e}_{\text{th}}(p_0, v_0)}, \tag{18.5}$$

$$\bar{Q} = \frac{\Gamma - 1}{\Gamma p_0 v_0}(H_0 - E^{0}) - 1. \tag{18.6}$$

Then we obtain some relations formally identical with those derived earlier for perfect gases, viz. (9.6), (10.2)–(10.4):

$$\frac{p}{p_0} = \frac{f + 1 + (2Q/p_0 v_0) - (v/v_0)}{(f + 1)\, v/v_0 - 1} = \frac{f + 1 + (f + 2)\,\bar{Q} - (v/v_0)}{(f + 1)\, v/v_0 - 1}, \tag{18.7}$$

$$f = \frac{2}{\Gamma - 1}$$

$$\frac{D_J^2}{\Gamma p_0 v_0} = \frac{v_0}{\Gamma p_0}\, \theta_J^2 = (\Gamma + 1)\,\bar{Q} + 1 + \sqrt{(\Gamma + 1)^2\, \bar{Q}^2 + 2(\Gamma + 1)\,\bar{Q}},$$

$$\frac{v_J}{v_0} = 1 + \bar{Q} - \bar{Q}\sqrt{1 + \frac{2}{(\Gamma + 1)\,\bar{Q}}}, \tag{18.8}$$

$$\frac{p_J}{p_0} = 1 + \Gamma \bar{Q}\left(1 + \sqrt{1 + \frac{2}{(\Gamma + 1)\,\bar{Q}}}\right).$$

On the other hand, it follows from the mass balance, momentum balance, and the CJ condition that

$$\frac{\varrho_J}{\varrho_0} = \frac{v_0}{v_J} = \frac{D_J}{a_J} = \frac{\Gamma + 1 - p_0/p}{\Gamma}. \tag{18.9}$$

While (18.7) is approximately valid along $\mathcal{H}_e$ in the vicinity of CJ, Equations (18.8) and (18.9) are rigorous with $\bar{Q}$ defined by (18.6).

Equations (10.1), (10.2) involving M_s remain valid only after M_s is replaced by a pseudo shock Mach number M_s' according to

$$M_s'^2 = \frac{D^2}{\Gamma p_0 v_0}. \tag{18.10}$$

As far as the practically equivalent inequalities

$$p_J \gg p_0, \quad Q \gg p_0 v_0 \tag{18.11}$$

hold, the important approximate equations (10.5) remain valid with Γ, M_s' instead of γ, M_s.

(18.11) is in general, like in Table 16.1, satisfied moderately for detonating gases and very accurately for condensed explosives. Since in the latter case in general $p_0 \leq 1$ atm and frequently even $p_0 \ll 1$ atm (explosions at high altitudes) p_0 has under these conditions no remarkable influence on the detonation process and is no longer an appropriate reference value. In this case we therefore set

$$p_0 = 0 \tag{18.12}$$

and have,

$$Q = H_0 - E^0 = E_0 - E,\circ \quad \frac{\varrho_J}{\varrho_0} = \frac{D_J}{a_J} = 1 + \Gamma^{-1}$$
$$p = 2(\Gamma - 1)\,\varrho_0 Q, \quad D_J^2 = 2(\Gamma^2 - 1)\,Q \tag{18.13}$$

instead of the dimensionless relations (10.5) which become meaningless.

19 Equations of State for Real Fluids

For the thermal equation of state of fluids, gases, and liquids, there exists a number of formulas which have been proposed to describe the thermodynamic behavior for states where the deviations from a perfect gas are distinct. For a fairly comprehensive treatment reference is made to [45].

In most of the cases the improvement starts from the perfect gas concept as the limiting case for small pressures and large specific volums adapting the formulas according to the observed behavior by means of suitable corrections.

In the following we list some of the best known of those improved equations of state:

a) van der Waals (1873):

$$\left(p + \frac{a}{v^2}\right)(v - b) = nR_aT; \tag{19.1}$$

b) Virial equation (Kamerlingh Onnes 1901):

$$\frac{pv}{nR_aT} = 1 + \frac{B(T, n_i)}{v} + \frac{C(T, n_i)}{v^2} + \cdots \tag{19.2}$$

c) Abel (1875):

$$p(v - b) = nR_aT; \quad b = \sum_i n_i b_i; \tag{19.3}$$

d) Cook (1947):

$$p[v - b(v)] = nR_aT; \tag{19.4}$$

e) Jones (1947):

$$p[v - b(p)] = nR_aT; \tag{19.5}$$

f) Becker (1922)—Kistiakowsky–Wilson (1941) (BKW):

$$pv = nR_aT(1 + xe^{\beta x}),$$
$$x = \frac{K}{vT^{\alpha}}, \tag{19.6}$$
$$K = \varkappa \sum_i x_i b_i;$$

g) Boltzmann—van der Waals—Berger (1960):

$$p + \frac{a}{v^2} = \frac{nR_aT}{v}(1 + x + 0.625x^2 + 0.287x^3 + 0.193x^4),$$
$$n = \sum n_i; \quad x = \frac{1}{v}\sum n_i b_i; \tag{19.7}$$

h) Lennard–Jones—Devonshire (1937):

In this case a thermal equation of state is obtained by summing over all interactions between the molecules starting with a simple assumption about the intermolecular potential and the idea that the molecule is enclosed in a cell which is determined by its neighboring molecules, see the References [45] p. 293 and [17] p. 82.

The simplest way of taking into account the real gas effects, viz. molecular attraction and co-volume of the molecules, is represented by Eq. (19.1) involving the constants a and b.

Equation (19.2) is ver ygeneral and in principle exactly valid. The convergence of the series, however, is guaranteed only for moderately dense gases. For the high densities in which we are interested here the formula is practically of no use.

Equation (19.3) as simplification of (19.1) is in so far admissible for the reaction products of gun powder and explosives as mostly high pressures are considered neglecting condensation phenomena. The molecular attraction taken into account by the constant a in (19.1) has then practically no effect. The *co-volume* b includes besides the volume of the gas molecules also the volume of the condensed components which are assumed to be incompressible. It has been found, however, that the assumption of a constant co-volume was too inaccurate. The assumption $b \approx$ const remains valid for low densities or pressures (see Fig. 19.1). It becomes, however, wrong for high densities especially because always $b < v$ must hold.

Equations (19.4) and (19.5) represent a considerable improvement as compared with the former equations. They, however, do not account for the particular composition of the gas and are unsatisfactory in this respect.

Equations (19.6) and (19.7) are much better regarding the effects of a particular composition and have been used successfully for the calculation of CJ states in Refs. [68] and [72].

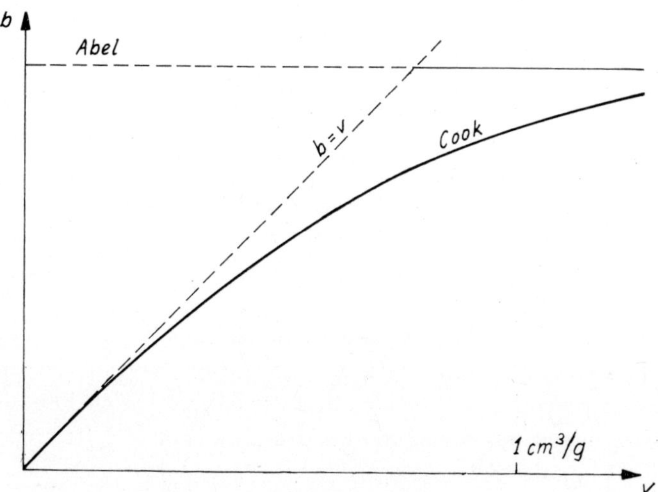

Fig. 19.1 Sketch describing the different assumptions made about the co-volume of the molecules in the equations of state of Abel and Cook.

If as in this case several parameters are available to adjust the equation to the empirical data the usefulness and accuracy of an equation of state is no longer necessarily determined by its form and its physical motivation. The establishment of a general equation of state with no arbitrary parameters, valid for all mixtures and states comprising particularly nonequilibrium states, even if only for C–H–O–N compounds, seems to be hopeless. The interactions between the numerous species ($k \geq 12$, see Section 16) are very complicated for high densities and temperatures and cannot be mastered theoretically. The ∞^{k+1} states are also experimentally inaccessible due to their manifold and frequent instability.

Since the measurements mainly comprise mechanical quantities at thermochemical equilibrium states only, they can hardly be used for proving the selected thermal equation of state. It was e.g. possible already for a long time, when sufficient accurate measurements of the detonation pressure were not available, to quantitatively explain the empirically observed dependence of the detonation velocity on the density of pressurized explosives by means of various thermal arrangements. The uncertainty of the thermodynamic assumptions became then particularly apparent in the resulting values for the CJ temperature; see Fig. 19.2.

If measurements of the CJ pressure are taken into account the uncertainty in selecting the proper thermal equation of state is considerably reduced. But still a strong discrepancy for T_{cJ} is found as well between the values derived from the different theories as compared with the few available experimental results, [68].

In Ref. [68] also the equation of state resulting from the Lennard–Jones–Devonshire free-volume theory has been used for calculating detonation processes since particularly in the high pressure range a better accuracy was expected. However, no essential improvement was found.

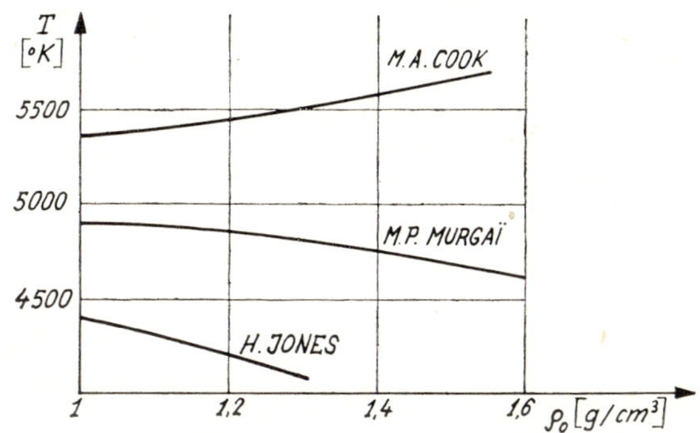

Fig. 19.2 Temperature behind a detonation wave in PETN calculated by means of different equations of state from the density-dependent detonation velocity (from Ref. [53]).

The above mentioned difficult and rather unsatisfactory methods which start from a thermal equation of state seem to be indispensable if a fairly complete knowledge of the occurring physico-chemical states in the reaction products is desired. Such a comprehensive description is particularly required as a basis for chemical-kinetic calculations.

If we however abandon to account for the nonequilibrium states and the chemical and thermocaloric quantities the mechanical equation of state for the reaction products can then be determined empirically [70], [71]. For normal reflection of the detonation wave in a solid explosive (here composition *B*, compare Table 16.1) at a number of inert materials with different densities and known mechanical properties a corresponding number of different states

of motion is generated which may be plotted in an u–p diagram, see Fig. 19.3; compare also Section 24.

The material velocity u can be measured. The pressure p can then be derived from the value of u by means of the Hugoniot curve of the inert material. On the u–p curve determined in this way particularly the CJ point will be found. It can be obtained as the intersection point with the line

$$p = \varrho_0 Du$$

according to (3.6). The branches of the curve above and below the CJ point correspond to the shock compression and isentropic expansion of the detonation products respectively. They can be converted to p–v curves (shock adiabat and reversible adiabat) by means of simple formulas, see Section 24. In [71] it became apparent that the polytrope

$$p = \text{const} \cdot \varrho^{\gamma} \quad (\gamma = \text{const}), \tag{19.8}$$

with $\gamma = 2.77$ in this case is a good approximation of the isentrope for higher pressures. For pressures $p_{cJ} > p > 570$ atm an improved empirical formula compared with (19.8) is given in the cited Reference.

Repetition of the experiments with different initial densities ϱ_0 leads to different entropy values S_{cJ} and therefore different isentropes. This permits to measure mechanically a two-dimensional region of equilibrium states for

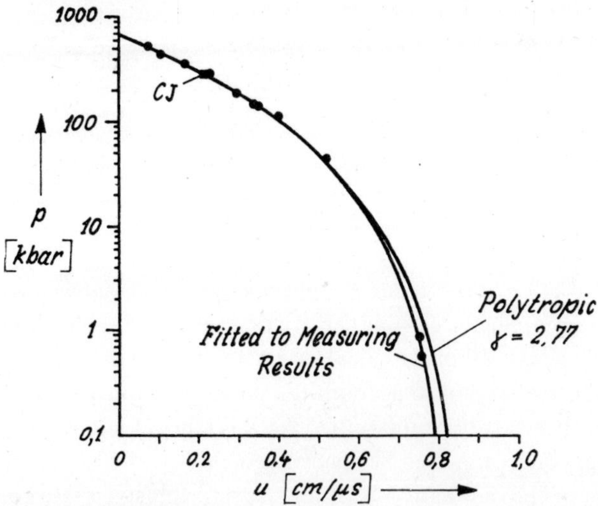

Fig. 19.3 Experimentally obtained adiabat for the reaction products of Grade A composition B (from Ref. [71]).

the reaction products of the detonation. The mechanical equation of state obtained in this way is basically necessary and sufficient for the gasdynamic calculation of the flow processes induced by the detonation.

Another method for the determination of the equation of state of the detonation products, which is feasible in principle even though practically of limited application, is the so-called *inverse method* or variation method.

This method, developed for the indirect determination of the detonation pressure in the CJ plane, in principle also permits to determine the relation between p, v and H and thereby the desired mechanical equation of state. The detonation pressure is determined from the different detonation velocities measured for various initial densities and energies which cause a variation of the CJ point. For a more detailed outline see Refs. [65], [66].

20 Chemical Equilibrium Calculation

In order to describe the physico–chemical behavior of the equilibrium states of a reacting mixture it is necessary to calculate the equilibrium composition of the substance for given thermodynamic conditions. For this purpose we ought to know a complete equation of state corresponding to the information level 2 in Section 4.

The procedure of an equilibrium calculation can be discussed here only briefly. For a more detailed discussion reference is made to the literature, particularly the text books [73], [14], [76], the conference proceedings [77], and Ref. [78] which is especially related to detonation states.

Suppose n_i ($i = 1, \dots, k$) are the numbers of moles of the k components or species X_i found in the unit mass of the mixture. The respective state of the system is determined by means of the numbers n_i and two thermodynamic variables, e.g. p and T. Chemical equilibrium is then obtained if for given p and T the *free enthalpy* G has a *minimum value*; see definition (2.10).

We therefore need an equation of state

$$G = G(p, T, n_i) \tag{20.1}$$

and are looking for the values n_i which for fixed p and T yield a minimum of G. It is, compare (20.26) and [14],

$$dG = v\,dp - S\,dT + \sum \mu_i\,dn_i. \tag{20.2}$$

The quantities

$$\mu_i = \left(\frac{\partial G}{\partial n_i}\right)_{p,T} \tag{20.3}$$

are called the *chemical potentials* of the components X_i. The numbers n_i may change due to chemical reactions

$$\sum_{i=1}^{k} \nu_i^j X_i = 0 \quad (j = 1, \ldots, l), \tag{20.4}$$

(j is an index!).

The reaction equations (20.4) are written in such a form, that the integer *stoichiometric coefficients* ν_i^j of the i-th component in the j-th reaction are set negative for the initial components and positive for the reaction products. For example

$$2H_2O - 2H_2 - O_2 = 0.$$

If we permit only linearly independent reaction equations, their number is always

$$l < k.$$

Namely, since the gross composition of the substance is prescribed and constant, the n_i have to satisfy in addition a number of linear equations with constant coefficients

$$\sum_{i=1}^{k} \lambda_i^{j'} n_i = c_{j'}, \tag{20.5}$$

which express the invariability of the total amount of the various atomic species and the electric charges. If e.g. C, H, O, N are participating in the reaction and ions occur, we have to satisfy five balance equations (20.5) and it is $l = k - 5$.

If we admit l reactions (20.7) for the k components, so that after introducing l reaction variables ξ_j the molar numbers are changing according to

$$dn_i = \sum_{j=1}^{l} \nu_i^j \, d\xi_j \tag{20.6}$$

we consequently obtain $k - l$ independent linear combinations of the n_i which remain invariant. Therefore, we may supplement (20.5) by

$$j' = l + 1, \ldots, k. \tag{20.5'}$$

For the vectors ν^j and λ^j an orthogonality relation exists

$$\sum_{i=1}^{k} \nu_i^j \lambda_i^{j'} = 0 \quad (0 < j \leqq l < j' \leqq k).$$

At the minimum of (20.1) we necessarily have

$$\sum \mu_i \, dn_i = 0$$

as long as the n_i are changing according to (20.6). This means

$$\sum_{i=1}^{k} \nu_i^j \mu_i = 0 \quad (j = 1, \ldots, l) \tag{20.7}$$

must hold. The l equilibrium conditions (20.7) together with the $(k - l)$ balance equations (20.5) determine the equilibrium position in general uniquely.

In principle, by using modern numerical computation techniques, either the last mentioned system of equations can be solved or the minimum problem together with the restricting conditions can be attacked directly. Both methods are discussed extensively in a recent work [79], certainly for the case of a homogeneous perfect gas phase only.

The quantity derived from (20.7)

$$A_j = -\frac{1}{T} \sum_i \nu_i^j \mu_i \tag{20.8}$$

is termed *affinity* in [59], and may be interpreted as a chemical driving force. A_j has always the same sign as the reaction rate $r_j = d\xi_j/dt$, and for $E =$ const, $v =$ const the entropy production or the measure of irreversibility is described by

$$\frac{dS}{dt} = \sum A_j r_j .$$

The equilibrium condition (20.7) can also be written as

$$A_j = 0 \quad (j = 1, \ldots, l).$$

In general, the state function (20.1) is not known explicitly. Suppose, however, tentatively that $G(p, T, n_i)$ is known for a fixed value p_1 of p and all T and n_i and that in addition the thermal equation of state

$$v = v(p, T, n_i) \tag{20.9}$$

is given for the whole range of variables. Then, according to (20.2)

$$v = (\partial G/\partial p)_{T, n_i}$$

and

$$G(p, T, n_i) = G(p_1, T, n_i) + \int_{\substack{p_1 \\ (T,\, n_i = \text{const})}}^{p} v(p, T, n_i)\, dp .$$

Now indeed $G(p_1, T, n_i)$ is quite well known for small values of p_1, namely, for the perfect gas state. Therefore, furthermore only Eq. (20.9) is decisive. The importance of the perfect gas mixtures on the one hand and the thermal equation of state on the other hand for the calculation of a chemical equilibrium in the general case is based on this fact.

The relations valid for a perfect gas mixture are briefly listed in the following. The quantities given herein have the meaning:

$$n = \sum_{i=1}^{k} n_i \tag{20.10}$$

is the specific total number of moles,

$$x_i = \frac{n_i}{n} \tag{20.11}$$

are the mole fractions,

$$p_i = x_i p \tag{20.12}$$

the partial pressures, and

$$c_{pi}(T), \quad H_i(T), \quad S_i^0(T), \quad \mu_i^0(T)$$

are the corresponding thermo-chemical quantities of the single components. S_i^0 and $\mu_i^0(T) = G_i(p_0, T)$ are values with $p = p_0 = 1$ atm as reference pressure. It is

$$\begin{aligned} \mu_i^0 &= H_i - TS_i^0, \quad S_i^0 = -\frac{d\mu_i^0}{dT}, \\ c_{pi} &= \frac{dH_i}{dT} = T\frac{dS_i^0}{dT}. \end{aligned} \tag{20.13}$$

Values of $c_{pi}(T)$, $H_i(T)$ and $S_i^0(T)$ are tabulated for many substances e.g. in [23].

H, S, G are additive for a perfect mixture according to

$$H = \sum n_i H_i, \quad S = \sum n_i S_i, \quad G = \sum n_i G_i; \tag{20.14}$$

for c_p the summation gives the frozen value

$$c_{pf} = \sum n_i c_{pi}. \tag{20.15}$$

The perfect gas equation (2.15) writes in this case

$$v(p, T, n_i) = \frac{nR_aT}{p} \tag{20.16}$$

with n from (20.10). The portions S_i, G_i of the components in the mixture, or of the components alone at the pressure p_i, are

$$S_i = S_i^0 - R_a \ln \frac{p_i}{p_0}, \tag{20.17}$$

$$G_i = \mu_i = \mu_i^0 + R_a T \ln \frac{p_i}{p_0}. \tag{20.18}$$

For the mixture we are able to distinguish between *mixing contribution* and *pressure contribution*

$$\ln \frac{p_i}{p_0} = \ln x_i + \ln \frac{p}{p_0}.$$

The equilibrium condition (20.7) takes now owing to Eq. (20.18) the form

$$\sum \nu_i^j \ln \frac{p_i}{p_0} = -\frac{1}{R_a T} \sum \nu_i^j \mu_i^0 = \frac{A_j^0}{R_a},$$

$$\prod_{i=1}^{k} \left(\frac{p_i}{p_0}\right)^{\nu_i{}^j} = \exp\,(A_j^0/R_a),$$

$$\prod_{i=1}^{k} p_i^{\nu_i{}^j} = \exp\,(A_j^0/R_a) \cdot p_0^{\nu^j} \equiv K_j(T). \tag{20.19}$$

Here, defined in analogy to (20.8), $A_j^0 - -(1/T)\Sigma\nu_i^j\mu_i^0$ is a *normal affinity*, only dependent on T. Furthermore is

$$\nu^j = \sum_i \nu_i^j = \frac{\partial n}{\partial \xi_j} \tag{20.20}$$

the increase of the mole number for a formula conversion of the j-th reaction, and $K_j(T)$ the equilibrium constant of this reaction. Formula (20.19) is one form of the mass action law (Guldberg and Waage, 1867). The $K_j(T)$ are tabulated for many reactions in [23].

Resulting from Eqs. (20.14), (20.18), (20.6) another interpretation for the affinity is obtained. It is

$$\sum \nu_i^j \mu_i = -TA_j = \frac{\partial G}{\partial \xi_j} = \sum \nu_i^j G_i = \Delta_j G$$

the increase of the free enthalpy per formula conversion of the reaction number j. In addition the following relations hold

$$\ln (K_j \cdot p^{-\nu^j}) = -\frac{\Delta_j G^0}{R_a T},$$

$$\frac{\partial (G/T)}{\partial T} = -\frac{H}{T^2},$$

$$\frac{d \ln K_j}{dT} = \frac{\Delta_j H}{R_a T^2} = \frac{Q_{pT}^j}{R_a T^2}, \tag{20.21}$$

$$\Delta_j H = \sum \nu_i^j H_i = \left(\frac{\partial H}{\partial \xi_j}\right)_{p,T} = Q_{pT}^j = q_{pT}^j(T). \tag{20.22}$$

The latter gives the amount of heat to be added per formula conversion; compare (5.3), (5.10). Also this quantity can be found tabulated in [23].

It follows from (20.19), (20.21) that for endothermal reactions ($Q_{pT} > 0$) the normal affinity and the equilibrium constant are increasing with T. That means at high temperatures the forward reaction is dominating, or we may say the equilibrium is shifted to the right hand side. For exothermal reactions the reverse situation is true, for high temperatures a reversal occurs.

From (20.10), (20.11), (20.12) it follows that

$$\sum_i p_i = p. \tag{20.23}$$

Instead of (20.5) we may also write

$$n \sum_{i=1}^{k} \lambda_i^{j'} p_i = p c_{j'}, \quad (j' = l + 1, ..., k). \tag{20.24}$$

The calculation of the chemical equilibrium has now been reduced to the solution of an algebraic system of Equations (20.19), (20.23), (20.24) with $k + 1$ equations for $k + 1$ unknowns p_i and n. If v and T are given instead of p and T the mathematical problem is varied only slightly due to (2.15).

For a small number of unknowns or for a simple structure of the system of equations the classical methods of algebra are used to obtain the solution. For a greater number of unknowns, in general, part of the p_i can be expected to be small, $p_i \ll p$, and iterative approximation methods may then be applied. This should be applicable and sufficiently easy in almost all cases of pure C–H–O–N mixtures. For a great number k and little knowledge about the character of the solution in advance we depend on the somewhat cumbersome methods described in [79].

In the case of a real single-phase fluid mixture with continuous transition into the perfect gas state for $p \to 0$ it is practical to associate with each state point $\{p, T, n_i\}$ besides the real state a fictitious perfect gas state as well as two corresponding functions G and G^{id}. It is then

$$\lim_{p \to 0} (G(p, T, n_i) - G^{id}(p, T, n_i)) = 0,$$

and G^{id}, v^{id}, μ_i^{id} are governed by the above relations. Particularly according to (20.16), (20.10)

$$\begin{aligned} v^{id} &= \frac{nR_aT}{p} = \frac{R_aT}{p} \sum_i n_i, \\ \frac{\partial v^{id}}{\partial n_i} &= \frac{R_aT}{p}. \end{aligned} \tag{20.25}$$

The real quantities are obtained from the perfect quantities by means of a correction:

$$G(p, T, n_i) = G^{id} + \int_0^p (v - v^{id})\, dp. \tag{20.26}$$

In the cases of interest to us we have always

$$\begin{aligned} v &> v^{id}, \\ \frac{\partial v}{\partial n_i} &> \frac{\partial v^{id}}{\partial n_i}, \end{aligned} \tag{20.27}$$

that means, each component requires due to repulsive forces actually more space than in the perfect gas state. From Eqs. (20.3), (20.18), (20.26) it follows now that

$$\mu_i = \mu_i^{0,id}(T) + R_aT \ln \frac{p_i}{p_0} + \int_0^p \frac{\partial (v - v^{id})}{\partial n_i}\, dp \tag{20.28}$$

with p_i according to the former definition (20.12), (20.11). In analogy to (20.18) relation (20.28) can be written in the form

$$\mu_i = \mu_i^{0,id} + R_aT \ln \frac{p_i^*}{p_0} \tag{20.29}$$

if the partial *fugacities* p_i^* are defined by (compare also (20.25)):

$$\ln \frac{p_i^*}{p_i} = \int_0^p \left(\frac{1}{R_a T} \frac{\partial v}{\partial n_i} - \frac{1}{p} \right) dp \quad (i = 1, \ldots, k). \tag{20.30}$$

Resulting from (20.27) it is $p_i^* > p_i$ and at our extreme conditions mostly even $p_i^* \gg p_i$.

By means of Eq. (20.29) the equilibrium conditions (20.7) can be written in a form analogous to (20.19) as

$$\prod_{i=1}^{k} (p_i^*)^{\nu_i^j} = K_j(T) \quad (j = 1, \ldots, l). \tag{20.31}$$

The equilibrium calculation for the real gas mixture is therefore now performed by means of Eqs. (20.23), (20.24), (20.30), (20.31).

In order to simplify, the expression (20.30) is often assumed to be independent of i and a *fugacity coefficient* k_f is defined by

$$p_i^* = k_f \cdot p_i \quad (k_f > 1).$$

Regarding (20.20) Eq. (20.31) takes now the form

$$\prod_{i=1}^{k} p_i^{\nu_i^j} = K_j(T) \cdot k_f^{\nu^j},$$

hereby retaining the partial pressures as compared to (20.24), but instead varying the equilibrium constants for those reactions which change the total mole number.

In the case of a system with arbitrary many phases for which the chemical-thermodynamic equilibrium is to be determined the entire formalism becomes more complicated but without adding new ideas or laws. The general case is discussed extensively in Ref. [80].

The fresh detonation products of C–H–O–N explosives often contain besides the fluid phase solid carbon in form of fine distributed soot, so that a two-phase equilibrium has to be calculated. The complicated procedure of [80] is not necessary in this case. Nevertheless the system of equations to be solved has a different form in the two-phase case than in the single-phase case. This problem is discussed in more detail in Ref. [53], p. 186.

For a considerable lack of oxygen the amount of O is not sufficient to convert all C and H atoms into CO and H_2O molecules and therefore always C_{solid} occurs in those cases; never, however, if sufficient O is available to form even H_2O and CO_2. In between both situations are possible. In the case of doubt only the equilibrium calculation shows whether the second phase is present or not. High pressure supports the deposit of soot as a reaction with decreasing gas molar number. Therefore, for pressed Nitropenta according to [53] C_{solid} appears in the CJ state exactly then, when the density of the explosive exceeds 1.2 g/cm³. The discontinuity between the single-phase and two-phase region becomes apparent in a sharp bend of the $\dot{D}$ versus ϱ_0 curve which has also been proved experimentally; see Fig. 20.1 which is reproduced from [53].

Other condensed phases may occur e.g. for explosives with metal compounds by forming solid metal oxydes; see Ref. [68]. In an extensive adiabatic expansion of the reaction products also water is formed as liquid phase as has been mentioned already in Section 17.

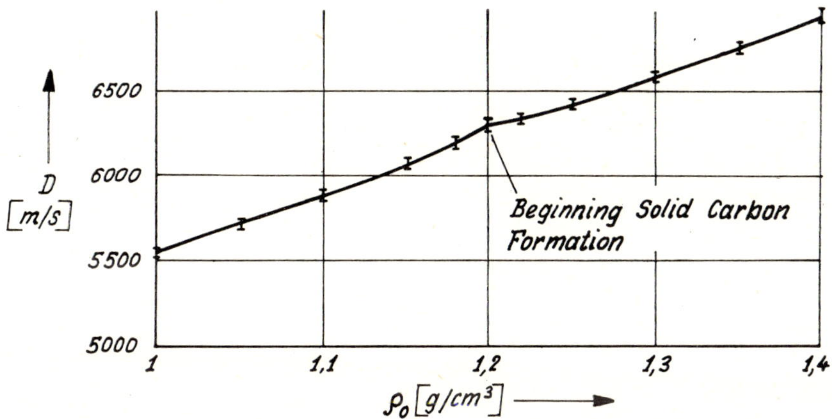

Fig. 20.1 Detonation velocity of PETN with 2% paraffin; from Berger and Viard [53] p. 190.

CHAPTER VII

Similarity Solutions

IN THE FOLLOWING CHAPTERS we shall discuss flow phenomena which are directly caused by a detonation process. The discussion will be restricted to simple processes which are sufficiently obvious for a theoretical and mathematical description.

At the very high pressures produced by detonations also solid materials behave almost like fluids so that the discussed flow phenomena can be considered as very general.

As before an ideal detonation is assumed in Chapters VII and VIII; this means:

a) Behind the detonation front exists the CJ state, provided the detonation is not overdriven.

b) The thickness of the reaction zone is negligibly small. The detonation wave similarly to an ordinary shock wave is considered as a mathematical discontinuity.

21 **The Similarity Law of C. Cranz**

21.1 *Similarity*

The important idea of physical *similarity* is based on the *dimensional analysis* which itself is closely related to the physical *system of units*. All three fields are extensively described in the literature; for similarity theory see [81], [82].

For our purpose we may select as *basic quantities* the length (l), the mass (m), the time (t), the temperature (T) and the molar mass (M_{mol}). Their *units*, which are the basic units of the selected system, are, as is well known, prescribed independently from each other by convention. The *dimensions* and units of all other derived quantities follow from the dimensions and units of the basic quantities as power products.

As long as the basic quantities of the considered processes are physically really independent, that means not formally related, a model-type correspondence between *similar* but not identical processes is possible. Each basic dimension allows of an arbitrary scale factor. From these the scale factors of the derived quantities result as power products corresponding to their dimensional formulas.

Each *dimensional constant* fundamental for the process restricts, however, the independent variability of the basic quantities. The *universal gas constant* e.g., with the dimension

$$[\mathcal{R}] = [l^2 t^{-2} T^{-1} M_{\text{mol}}],$$

has the same value for similar processes and reduces the number of independent basic quantities from five to four. Correspondingly in Ref. [84] a derived dimension

$$[M_{\text{mol}}] = [l^{-2} t^2 T]$$

was assigned to the molar mass. In blasts and explosions in general certain inactive substances are participating with characteristics influencing essentially the process. Their material quantities must be invariant for similarity transformations resulting in further dimensional constants. These are particularly *pressures* as detonation pressure or the reciprocal of the elasticity modulus, and *densities* as initial densities of condensed substances. Due to these two additional invariance conditions only two instead of four independent basic quantities are left. Resulting from the dimensional relations $[p/\varrho] = [u^2] = [E]$ together with the pressure and density implicitly also the velocity (u) and the specific energy (E) are similarity-invariant.

The *temperature*, so far still considered as free variable and anyhow being not an essential quantity for the predominating mechanical processes, is related to the pressure and the density by means of the equation of state (or the material-specific, invariant molar masses) and therefore also invariant. Consequently only one independent basic dimension is left which can be used for similarity relations. We can select (l) as this basic dimension. The fact, that no characteristic constant length exists for the particular process under consideration is essential in this case for the possibility of similarity correspondences.

21.2 *Cranz' similarity law*

Resulting from these considerations a certain *similarity law* is found with a single arbitrary scale factor. It is named here after C. Cranz (1926), but sometimes attributed also to Hilliar (1919) and Hopkinson (1916), [84].

We make the following definition:

Two physical processes are similar in the sense of Cranz *if the spatial configurations are geometrically similar, that means if for corresponding lengths l of* *the one process (main process) and $\bar{l}$ of the other process (model process)*

$$\bar{l}/l = k = \text{const} \tag{21.1}$$

holds, if further for the times

$$\bar{t}/t = k, \tag{21.2}$$

and for the masses and energies

$$\bar{m}/m = \bar{E}/E = k^3 \tag{21.3}$$

holds, while the intensive thermodynamic quantities remain unchanged:

$$\bar{p}/p = \bar{\varrho}/\varrho = \bar{T}/T = 1. \tag{21.4}$$

According to the above discussions the Cranz similarity is extensively applicable for detonations as it is in general for gasdynamic processes. That means the transfer of the model results (for a generally small model test) to the similar original process is possible. Formulation and proof of this statement together with a thorough discussion of its limitations are given in [17], p. 240–244.

The validity of this similarity law becomes questionable or will be violated always then, if phenomena become effective which in connection with the already mentioned dimensional constants define a characteristic length. These may be chemical reactions (reaction zone length!), viscosity and ther-

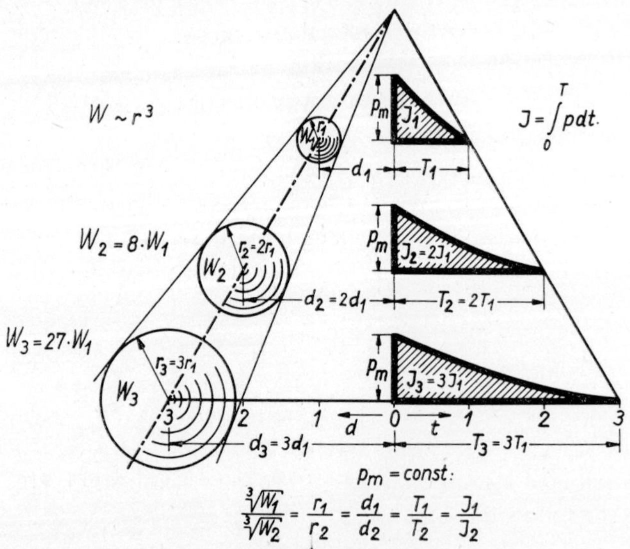

Fig. 21.1 Application of the *Cranz similarity law* to the characteristic quantities and measured values of a detonating explosive sphere.

mal conduction (boundary layer thickness), radiation, turbulence, relaxation, diffusion, plastic deformation, surface tension, gravity forces, grain size effects in microcristalline materials, etc.

Cranz' model law is like other similar laws an approximation. Its validity has to be proved theoretically or experimentally for each case. Shock fronts and detonation fronts are governed by the Cranz similarity law in so far, as their depth can be neglected compared with the dimensions of the experimental arrangement. This is no longer true for non-ideal detonations (Chapter IX).

The elastic behavior of solid bodies determined by the functional dependence of the stress tensor on the deformation tensor (not, however, on the deformation rate) is governed also by Cranz' law.

For the similarity transformation all non-dimensional quantities essential for the process remain of course invariant; particularly here all Mach numbers including the shock Mach numbers.

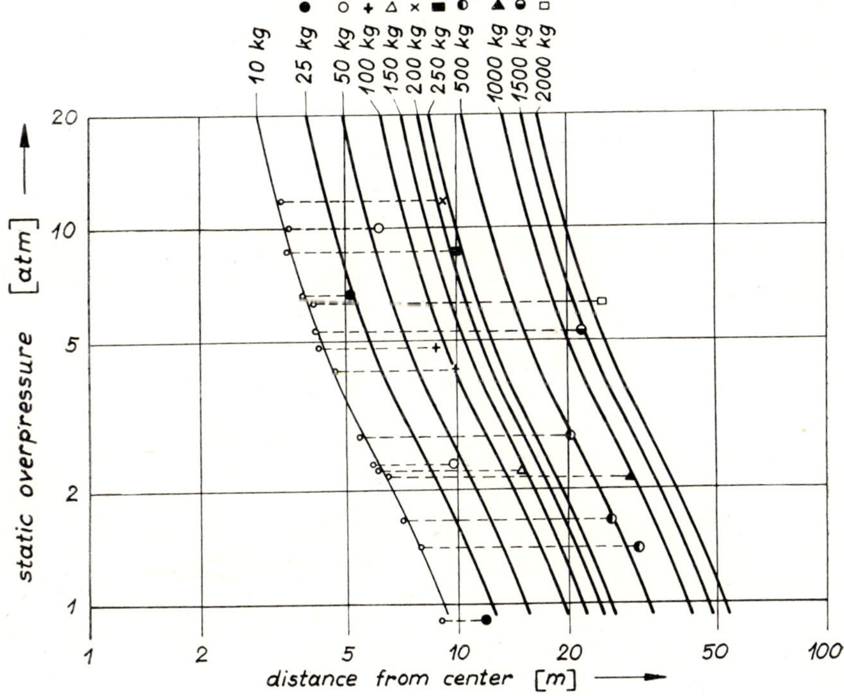

Fig. 21.2 Reduction of blast wave pressure measurements according to the *Cranz similarity law* to a normalized weight of charge: thin line.

Figures 21.1 and 21.2 show the application of the Cranz law to explosion processes in case of a center-initiated explosive sphere without casing.

In Fig. 21.1, the original sphere and two spheres, linearly enlarged according to the scale $k = 2$ and $k = 3$, respectively, are shown together with the corresponding schematically drawn blast pressure signatures measured at corresponding distances d_1, d_2, d_3. The spherical radii, the distances and the periods T_1, T_2, T_3 of the positive pressure phase are proportional to k. Consequently the same is true also for the impulses, I_1, I_2, I_3, the maximum pressure p_m being invariant. The charge weights W_1, W_2, W_3 are proportional to k^3.

Figure 21.2 shows the pressure maxima as function of the distance for various explosive charges. For each of the curves, however, only a few measuring points exist. According to Cranz' similarity law with $W/d^3 =$ const and $p_m =$ const the measurements have been reduced to a unit charge of 10 kg. Transformation of these values and redrawing yields then after smoothing out irregularities a reliable standard curve (thin line in Fig. 21.2). From this curve the peak-pressure curves for the different charge weights are obtained simply by horizontal parallel displacement in this double-logarithmic diagram.

22 Selfsimilar, Pseudostationary Flow Processes

If a physical process is a) governed by a certain one-parameter similarity law —therefore permitting similar magnification and diminution—, and b) reproduced identically by a diminution with a certain factor k, then such a process is called *selfsimilar** with respect to this law. Under reasonable conditions any real number can then be taken for k. That means the process is invariant with respect to a continuous, one-parameter group of similarity transformations.

Since in this case, similarly as for a geometrical symmetry group, the number of coordinates is reduced by one the mathematical description is in general simplified considerably. Early investigations of selfsimilar flow processes were done by BECHERT (1941) and GUDERLEY (1942). Selfsimilar processes particularly in the sense of C. CRANZ are discussed in [85] and termed *pseudostationary*. Those processes will be considered also in this text. In [86] general selfsimilar processes are discussed in detail and systematically.

The head-on (normal) incidence of a plane shock wave of uniform strength

* Another, less common term for selfsimilar is *homologous*.

on a wedge placed parallel to the shock front is an example of a simple plane, essentially pseudostationary flow process*. In so far the phenomenon can be described in terms of the coordinates x/t and y/t. The process is shown in Fig. 22.1. The incident shock wave is partly reflected and partly diffracted by the wedge [87]. Figur 22.1 shows two shadowgraphs of the same phenomenon at different instants. At the second instant, 40 μs later, the configuration of

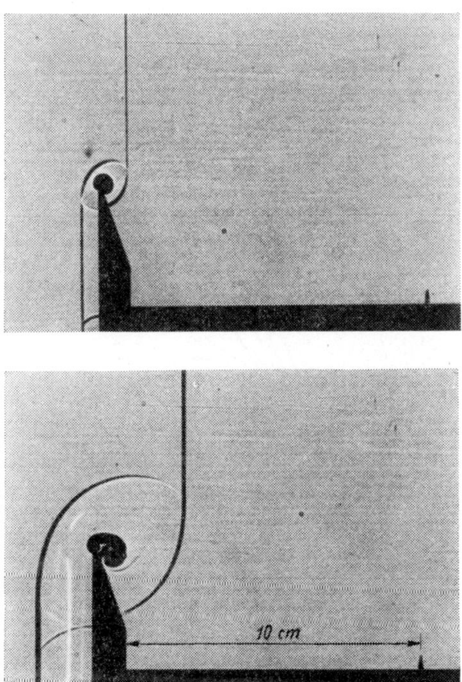

Fig. 22.1 Reflection and diffraction of a shock wave at a wedge (from Schardin). Two spark shadowgraphs with time interval 40 μs.

the shock and expansion wave system around the wedge is enlarged similarly by a factor $k = 3.2$ compared with the earlier configuration. However, also the limitations of the Cranz similarity law become apparent: The vortex generated at the wedge which separates gradually does not develop according to the similarity relation regarding its form and size.

Since the effect of the vortex on the pressure wave propagation is very small the wave pattern developes fairly accurately as selfsimilar process,

* This has been in principle realized already in 1943 by J. v. NEUMANN.

corresponding to a nonviscous flow without vortex formation. This behavior is described quantitatively in Fig. 22.2 by means of the space-time curves *a–l*, which are straight lines through the origin for the wave region (*g–l*), not however, for the vortex region (*a–f*).

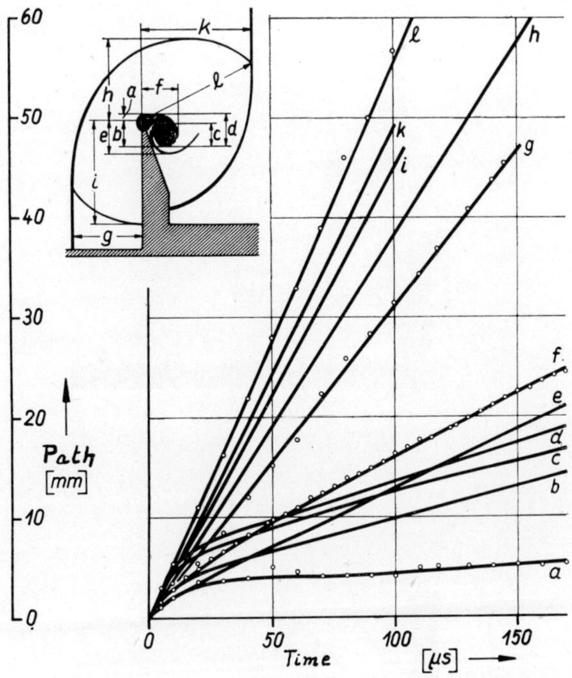

Fig. 22.2 Proof of the pseudostationary character of the wave process shown in Fig. 22.1. From Ref. [87].

With increasing time the viscous effects become less effective and the selfsimilarity increasingly accurate. The not selfsimilar boundary conditions, as e.g. the wedge base and walls, however, eventually terminate the selfsimilar property of the flow.

23 Selfsimilar Detonation, the Taylor Wave

Suppose a spherical charge is initiated at its center by a small detonator. An almost *spherical* detonation wave is then generated propagating through the explosive. The disturbance of the spherical symmetry occurring in reality due to the nonspherical detonator and its supply chord can be kept very small.

Neglecting the length of the reaction zone and the detonator we may consider the explosion as an ideal detonation in a homogeneous explosive initiated at the point $r = 0$ at the time $t = 0$ and propagating in form of concentric spheres given by

$$R = Dt. \tag{23.1}$$

As long as the detonation front has not yet arrived at the outer boundary of the explosive charge the boundary has no influence on the process and we may consider the charge as universally unlimited. With these conditions the entire process is pseudostationary and regarding in addition its spherical symmetry it can be described in principle by functions of a single variable

$$z = \frac{r}{t}. \tag{23.2}$$

If the initiation at $t = 0$ is performed not at a single point but rather along a straight line, e.g. in form of a wire explosion, the detonation propagates as a symmetrical *cylinder wave**. This case is mathematically very similar to the spherical case. Both may be analyzed together if r and R in the cylindrical case are taken as the distance from the symmetry axis and a dimensional number ν is being regarded, see e.g. Refs. [2] and [8]. The number ν has the values 3 and 2 for the spherical case and the cylindrical case, respectively.

With $\nu = 1$ also the case of a planar initiation can be taken into account, which may be realized technically e.g. by the impact of a metal plate on an explosive surface; see Chapter IX. Here r or R is the distance from this plane and relations (23.1) and (23.2) remain formally unchanged and valid. While, however, for $\nu > 1$ always $r \geq 0$, $z \geq 0$, for $\nu = 1$ also $r < 0$ with $z < 0$ is possible and meaningful.

The cases $\nu = 1$ and $\nu = 3$ have been analyzed independently, by G.I. TAYLOR (1941), published (1950) [83], YA.B. ZELDOVICH (1942) [28] Section 23, and W. DÖRING (1944) [17]. Consequently the solution usually attributed to TAYLOR should be called Taylor–Zeldovich–Döring solution. Sometimes, regarding the physical meaning of the solution, it is called *Taylor Wave.*

Since each fluid particle of the reaction products generated at the detonation front is bound to the isentrope through the CJ point the entire flow of

---

* Only if the initiation occurs simultaneously for each point along the axis the wave front is parallel to the cylinder axis. If the initiation point (initiation head) propagates with a certain finite velocity ($> D$) along the axis the wave front is conic with the cone angle dependent on this velocity. This case too is pseudostationary and can be described in terms of a single independent variable.

the reaction products is *homentropic* (see page 14). Always

$$dp = a^2\, d\varrho. \tag{23.3}$$

Eliminating p by means of (23.3) we write the basic equations for a homentropic unsteady flow as

$$\frac{\partial \varrho}{\partial t} + u \frac{\partial \varrho}{\partial r} + \varrho \frac{\partial u}{\partial r} + (\nu - 1) \frac{u\varrho}{r} = 0, \tag{23.4}$$

$$\frac{\partial u}{\partial t} + u \frac{\partial u}{\partial r} + \frac{a^2}{\varrho} \frac{\partial \varrho}{\partial r} = 0 \tag{23.5}$$

with $\nu = 1$ plane, $\nu = 2$ cylindrical, and $\nu = 3$ spherical symmetry.

Replacing r by z according to Eq. (23.2) and regarding the assumed *pseudostationarity*, which requires that

$$\frac{\partial \varrho(z, t)}{\partial t} = \frac{\partial u(z, t)}{\partial t} = 0,$$

we obtain for (23.4) and (23.5)

$$(u - z) \frac{d\varrho}{dz} + \varrho \frac{du}{dz} + (\nu - 1) \frac{u\varrho}{z} = 0, \tag{23.6}$$

$$(u - z) \frac{du}{dz} + \frac{a^2}{\varrho} \frac{d\varrho}{dz} = 0. \tag{23.7}$$

In addition to these equations describing the motion, we need an isentropic equation of state $p = p(\varrho)$ or

$$a = a(\varrho), \tag{23.8}$$

e.g. for a calorically perfect gas

$$a/a_J = (\varrho/\varrho_J)^{(\gamma-1)/2}, \tag{23.9}$$

with the suffix J relating to the CJ state.

We are looking at first for the solution of Eqs. (23.6)–(23.8) in the intervall $0 \leq r \leq R$, or according to (23.1), (23.2) in

$$0 \leq z \leq D, \tag{23.10}$$

which satisfies the boundary conditions

$$\varrho(D) = \varrho_J, \quad u(D) = u_J, \tag{23.11}$$

$$u(0) = 0. \tag{23.12}$$

This corresponds to an initiation at a single point ($\nu = 3$), along a straight line ($\nu = 2$), and along a plane ($\nu = 1$). The initiation plane in the case $\nu = 1$ can be a symmetry plane within the unlimited explosive or a plane rigid boundary of an explosive semi-space, see Fig. 23.1.

Instead of assuming an unlimited explosive the symmetry properties can be conserved if we assume a rigid casing of cylindrical or prismatic form for $\nu = 1$, wedge form for $\nu = 2$, and conical form for $\nu = 3$. Condition (23.12) can be replaced later by other conditions in the case $\nu = 1$.

There exists a close analogy between the cases $\nu = 2$ and $\nu = 3$. The plane case $\nu = 1$, however, is obviously qualitatively different and mathematically simpler since (23.6) is reduced by one term.

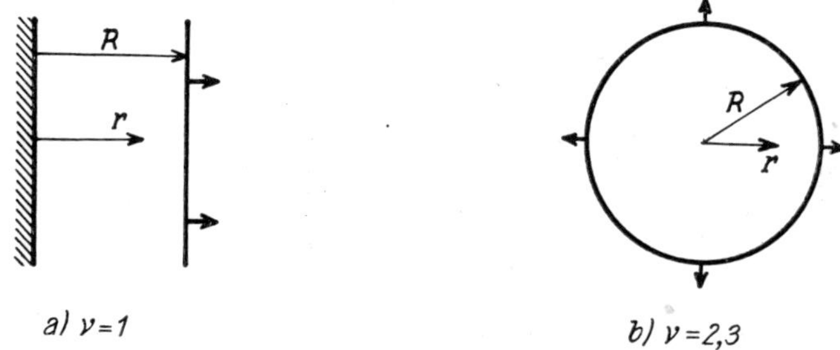

Fig. 23.1 The region $0 \leq r \leq R = Dt$ of the desired solution.

Equivalent to Eqs. (23.6), (23.7) is the symmetrical form

$$\left.\begin{aligned} (u - z + a)\left(\frac{a}{\varrho}\frac{d\varrho}{dz} + \frac{du}{dz}\right) + (\nu - 1)\frac{au}{z} = 0, \\ (u - z - a)\left(\frac{a}{\varrho}\frac{d\varrho}{dz} - \frac{du}{dz}\right) + (\nu - 1)\frac{au}{z} = 0. \end{aligned}\right\} \tag{23.13}$$

By introducing *Riemann's thermodynamic function*

$$y = \int_{\substack{0 \\ s=\text{const}}}^{\varrho} \frac{a}{\varrho}\, d\varrho = \int_{\substack{0 \\ s=\text{const}}}^{p} \frac{dp}{a\varrho} \tag{23.14}$$

which is defined along each isentrope Eqs. (23.13) can be written in the form

$$\left.\begin{aligned} (u - z + a)\frac{d(y + u)}{dz} + (\nu - 1)\frac{au}{z} = 0 \\ (u - z - a)\frac{d(y - u)}{dz} + (\nu - 1)\frac{au}{z} = 0 \end{aligned}\right\} \tag{23.15}$$

and (23.8) as

$$a = a(y). \tag{23.16}$$

For a perfect gas (23.16) becomes

$$a = \frac{\gamma - 1}{2} y. \tag{23.17}$$

The differential equations have of course the trivial solution

$$u = 0, \quad \varrho = \text{const}, \quad p = \text{const}, \quad a = \text{const}, \quad y = \text{const}. \tag{23.18}$$

This solution is valid not only in the undisturbed region $z > D$ but is also used as part of the actual solution region (23.10). Namely, since due to the twofold boundary condition (23.11) the nontrivial solution which starts at $z = D$ is already uniquely determined, there is no freedom left to satisfy also (23.12) by this solution. This, however, can be achieved by matching at the proper place

$$u = u_f = 0, \quad z = z_f, \tag{23.19}$$

the trivial solution (23.18), valid for $0 \leq z \leq z_f$, and the non-trivial solution, valid for $z_f \leq z \leq D$. Behind the actual Taylor wave, which is a rarefaction wave following the detonation front, a *constant pressure region* exists where the fluid is at rest.

For a more detailed discussion we have to consider the cases $\nu = 1$ and $\nu > 1$ separately.

For $\nu = 1$ according to (23.15)

$$(u - z + a)\frac{d(y + u)}{dz} = (u - z - a)\frac{d(y - u)}{dz} = 0. \tag{23.20}$$

At $z = z_J = D$ we have $z = u + a$ (CJ condition!) and $u - z - a \neq 0$. Therefore, with (23.20) we obtain for the desired solution

$$y - u = \text{const} = y_J - u_J. \tag{23.21}$$

Since, owing to (23.19), $y = \text{const}$ and $u = \text{const}$ do not hold individually it is $(d(y + u))/dz \neq 0$ and therefore

$$u - z + a = 0. \tag{23.22}$$

Equations (23.21), (23.22) together with (23.2), (23.14), (23.16) determine $u(z)$ and $y(z)$ and characterize the flow as a simple, centered, outward propagating, homentropic rarefaction wave. Condition (23.19) is supplemented by final values resulting from (23.21), (23.22)

$$y_f = y_J - u_J, \quad z_f = a_f = a(y_f). \tag{23.23}$$

Also $p(z)$, $\varrho(z)$, p_f and ϱ_f are found unique.

If $r = 0$ determines the explosive surface adjacent to *vacuum* instead of being a symmetry plane or rigid wall, (23.12) is to be replaced by $p = p_v = 0$ or

$$y_v = y\,(z_v) = 0 \tag{23.24}$$

and it results

$$a_v = 0, \quad u_v = z_v = -(y_J - u_J). \tag{23.25}$$

The solution (23.21), (23.22) terminates at the vacuum front (23.24), (23.25); the trivial solution has no application.

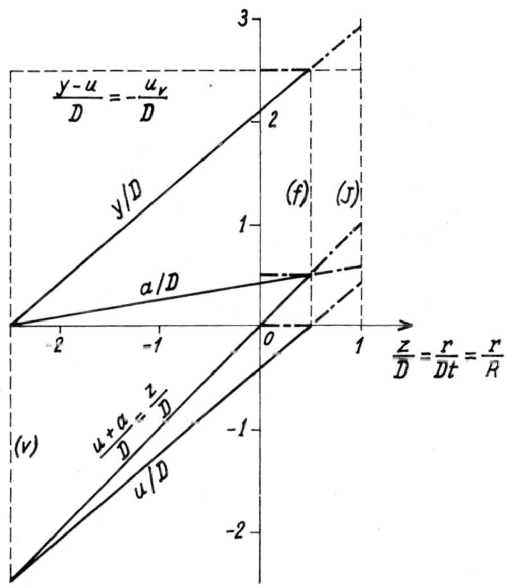

Fig. 23.2 Selfsimilar detonation ($\nu = 1, \gamma = 1.4$). (J): CJ state; (v): vacuum boundary; (f): final values; –·–·–: condition (23.12); ——: condition (23.24).

The case of more general rear boundary conditions of which (23.12) and (23.24) are special cases is discussed later in Chapter VIII.

For a calorically perfect gas with Eq. (23.9) and (23.17) u, a, and y become linear functions of z; see Fig. 23.2.

In the following we list some typical values for this case:

$$\left.\begin{aligned}
&z_J = D, \quad u_J = \frac{D}{\gamma + 1}, \quad y_J = \frac{2\gamma D}{\gamma^2 - 1}, \quad a_J = \frac{\gamma D}{\gamma + 1}, \\
&z_f = \frac{D}{2}, \quad u_f = 0, \quad y_f = \frac{D}{\gamma - 1}, \quad a_f = \frac{D}{2}, \\
&z_v = -\frac{D}{\gamma - 1}, \quad u_v = -\frac{D}{\gamma - 1}, \quad y_v = 0, \quad a_v = 0, \\
&\frac{p}{p_J} = \left(\frac{a}{a_J}\right)^{2\gamma/(\gamma-1)}, \quad \frac{\varrho}{\varrho_J} = \left(\frac{a}{a_J}\right)^{2/(\gamma-1)}.
\end{aligned}\right\} \tag{23.26}$$

In formulas (23.26) the approximations (18.11), (10.5) are used. Numerical values of the γ-dependent coefficients occurring in (23.26) are listed in Table 23.1.

Table 23.1 Coefficient values for formulas (23.26)

γ	3	5/3	7/5
$\frac{2}{\gamma - 1} = f$	1	3	5
$\frac{1}{\gamma + 1}$	1/4	3/8	5/12
$\frac{2\gamma}{\gamma^2 - 1}$	3/4	15/8	35/12
$\frac{\gamma}{\gamma + 1}$	3/4	5/8	7/12
$\frac{1}{\gamma - 1}$	1/2	3/2	5/2
$\frac{2\gamma}{\gamma - 1}$	3	5	7

For real explosives $u(z)$, $y(z)$, and $a(z)$ are in general not linear functions. Qualitatively, however, they have a similar development as that shown in Fig. (23.2).

Considering once more the cases $\nu > 1$ we resume our discussion starting again with (23.15). For the inner part of the actual Taylor wave, in the region

$z_f < z < D$, it is $dy/dz > 0$, $du/dz > 0$, and $(\nu - 1)\, au/z > 0$ and therefore

$$u - z + a < 0. \tag{23.27}$$

At $z = D$, however, this expression vanishes (CJ condition!) and

$$\left.\frac{d(y + u)}{dz}\right|_{z=D} = +\infty. \tag{23.28}$$

Since $d(y - u)/dz$ remains finite dy/dz and du/dz become infinite both in the same way*. This singularity can be overcome if we use u as independent variable. Equation (23.15) is equivalent to

$$\frac{dy}{du} = \frac{z - u}{a}, \quad \frac{dz}{du} = \frac{z}{u}\,\frac{(z - u)^2 - a^2}{(\nu - 1)\, a^2}. \tag{23.29}$$

The system (23.29) is to be integrated from the J values up to (23.19). It turns out that at the end of the integration intervall another, even more complicated singularity occurs. The inequality (23.27) becomes here the equality (23.22) resulting in a term 0/0 on the right hand side of (23.29). Eventually one finds that at this point $dz/du = \infty$ and consequently $du/dz = 0$, however, $d^2u/dz^2 = \infty$, see [88].

The numerical integration of the differential equations (23.29) can be performed relatively easy if an empirical adiabatic relation $p(\varrho)$ or $a(y)$ is given [83]. For a calorically perfect gas certain simplifications result but a numerical integration is still required [17]. Figure 23.3 shows the results from a calculation with $\nu = 3$ based on a semiempirical adiabatic relation which has been obtained from an Abel equation of state with constant co-volume and constant γ.

Before passing on it is suitable to add some further remarks regarding the relations (23.22) and (23.27). The directional condition for the rightward facing characteristics

$$\frac{dr}{dt} = u + a \tag{23.30}$$

determines the outward propagation of small pressure disturbances. Along the straight lines $r/t = z =$ const in the r–t plane (Fig. 23.4), which are known to us as lines of constant state, Equation (23.30) holds also due to (23.22) in the case $\nu = 1$. That means these rays are the lines determining the pressure disturbance propagation. Disturbances which are expected in any case as

* As to the physical implications hereof see Section 30.3.

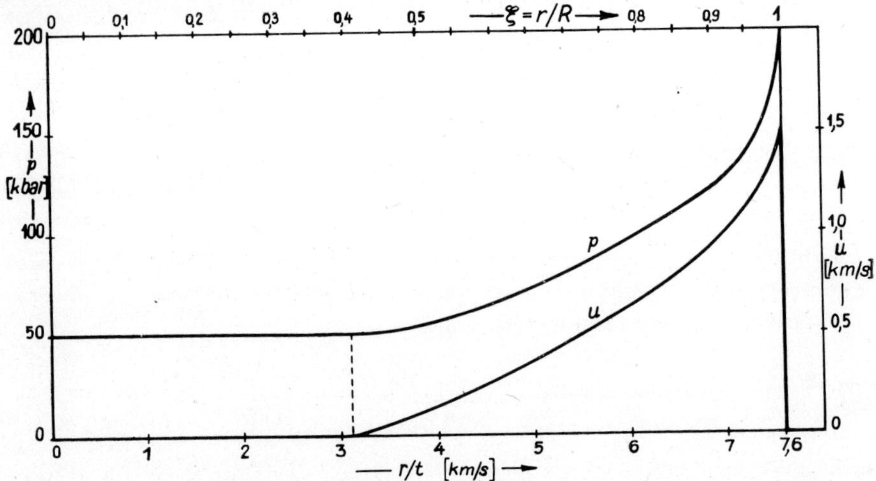

Fig. 23.3 Pressure and flow velocity for spherically symmetric pseudostationary detonation in TNT/RDX 50 : 50 (from Ref. [88]).

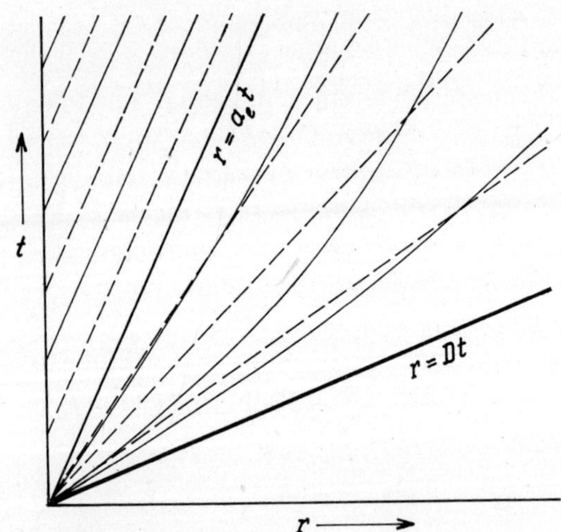

Fig. 23.4 Characteristics for the cases $\nu = 1$ (---) and $\nu > 1$ (—).

induced by the initiation process propagate in the case $\nu = 1$ along $z =$ const. Therefore, they are not shifted relatively to the Taylor wave. During the expansion process the disturbances become weaker and the theoretically derived ideal shape of the Taylor wave is approached asymptotically and becomes stable.

In the case $\nu > 1$ relation (23.22) is valid only at the head and the tail of the rarefaction wave. At intermediate positions (23.27) holds. That means the disturbances are propagating slower than with $z =$ const in the inner part of the Taylor wave. They have the tendency to accumulate at the tail of the wave ($z \approx z_f$) while being attenuated simultaneously.

For disturbances occurring in the constant pressure region ($z < a_f$) the same is anyhow valid, independently of ν: these disturbances propagate here with $dr/dt = a_f$, $r = a_f t +$ const, $z = a_f +$ const$/t$, so that for $t \to \infty$ here $z \to a_f$ holds. And indeed, disturbance impulses have been found experimentally at the tail of the Taylor wave*.

From (23.22) it further follows that each value $z =$ const $= z_1$ with $z_v < z_1 < D$ is permissible as separation plane between Taylor wave and constant pressure region and that further pseudostationary solutions with generalized boundary conditions result. This holds, of course, for $\nu = 1$ only, not for $\nu > 1$.

For the detonation of a finite charge, and especially for a spherical charge, the pseudostationary behavior ceases after the entire explosive has detonated. The theoretical-mathematical analysis of the following processes requires integration of partial differential equations, for which the initial conditions are obtained from the pseudostationary solution [88].

---

* According to a private communication by W. D. Struck.

CHAPTER VIII

Interaction of Detonations at Boundaries

IN THE ACTUAL CASE the detonating substance is always surrounded by an inert material. When the detonation front reaches the boundary surface both materials start to interact mechanically. In general a strong shock wave enters the inert material while a reflected pressure disturbance propagates into the detonation products. The practical interest in detonation processes, however, is mostly just in its mechanical action on its surroundings. We are therefore interested, at least to some extent, in a discussion of the resulting, frequently very complicated phenomena.

Starting as before with as simple as possible conditions we assume at first that the explosive and the inert material each occupy a semi-infinite space separated by a plane and that through the explosive a plane, ideal, quasi-steady CJ detonation is propagating. The inert material shall behave mechanically and thermodynamically as a fluid without anomalies. Therefore the acceleration of a metal casement of finite thickness is not treated here, although this process is basically important for explosive ballistics [53], [107].

Two schematic limiting cases of a highly compliant and a very rigid inert material are frequently considered: the vacuum and the perfectly rigid wall.

24 Normal Interaction

24.1 *The u–p diagram*

Suppose the explosive with pressure and density values p_0 and ϱ_0 and the inert material with p_0 and ϱ_0' are in contact at $x = 0$ so that the explosive occupies the semispace $x < 0$. Suppose further a detonation front $\mathscr{D}$ propagates in the explosive during $t < 0$ according to $x = Dt$ striking head on the boundary surface $\mathscr{B}$ at $t = 0$; see Figs. 24.1 and 24.2. The actual detonation is now terminated and the interaction between the reaction products with p_1, ϱ_1, u_1 and the inert material starts. The entire process is *selfsimilar* or pseudostationary and may be described by a single coordinate $z = x/t$.

There will be always a shock wave $\mathscr{S}'$ propagating with constant velocity S' into the second medium, the state 2′ behind the shock being homogeneous.

The reflected wave is a shock wave $\mathscr{S}$ or an expansion (rarefaction) wave $\mathscr{E}$ depending on whether the second medium is *hard* or *soft*. The homogeneous regions 2 and 2′ are in *mechanical equilibrium*:

$$u_2 = u_2', \quad p_2 = p_2'. \tag{24.1}$$

The exact values result from the initial conditions and the material properties according to a construction illustrated in Fig. 24.1c. The transmitted wave 0′–2′ and the reflected wave 1–2 are to be determined properly so that Eq. (24.1) is satisfied. That means, in each case two curves have to be brought to an intersection in a *u–p* diagram.

One of these curves ($\mathscr{S}'$) is the Hugoniot curve corresponding to the state 0′ which has been transformed to *u–p* coordinates. *u* is here the flow velocity

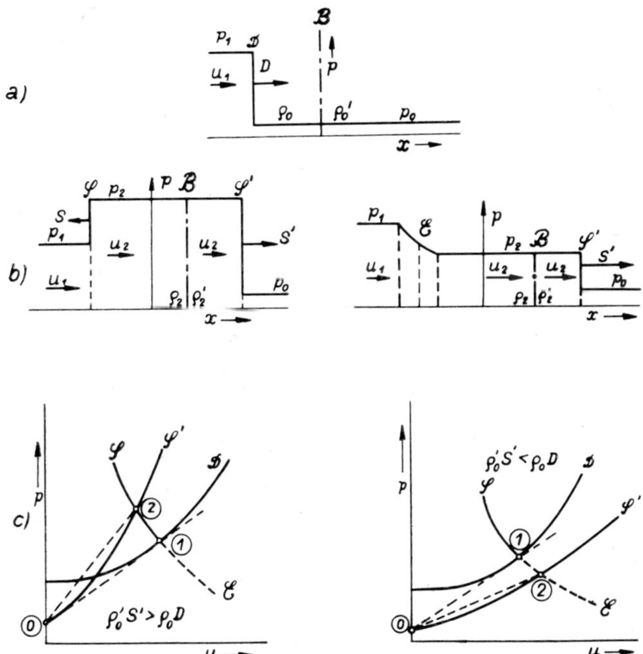

Fig. 24.1 Head-on interaction of a detonation wave with an inert material. a) before, b) after the collision. c) impedance construction with the hard reflection on the left, the soft reflection on the right side. $\mathscr{D}$ detonation front; $\mathscr{B}$ material boundary; $\mathscr{S}'$ shock produced in the second medium; $\mathscr{S}$ reflected shock; $\mathscr{E}$ reflected rarefaction wave.

behind the shock. Due to

$$\frac{\Delta p}{\Delta u} = \theta' = \varrho_0' S',$$

see Eq. (3.6), the *Rayleigh line* is a straight line too in the *u–p* diagram. It is shown as a dashed line in Fig. 24.1.

State 1 lies on the detonation Hugoniot curve $\mathscr{D}$ which belongs to the initial state 0. This state point is the CJ point and the Rayleigh line 0–1 is tangential to $\mathscr{D}$ at point 1. In the following only point 1 is required and not the entire curve $\mathscr{D}$.

The second geometrical locus for point 2 is the curve $\mathscr{S} + \mathscr{E}$ which corresponds to a shock or rarefaction wave propagating leftward into the explosion products originating from the initial state 1. $\mathscr{S}$ is again a transformed Hugoniot curve, but $\mathscr{E}$ a transformed isentrope. According to Eqs. (3.5), (3.6) with state 1 as reference state, along $\mathscr{E}$

$$(du)^2 = -dp\, dv$$

and along $\mathscr{S}$

$$(\Delta u)^2 = -\Delta p\, \Delta v.$$

The plane, pseudostationary rarefaction wave is already familiar to us from the discussion in Sec. 23.

It is quite evident that between the curves $\mathscr{S}'$ and $\mathscr{S} + \mathscr{E}$ in the *u–p* diagram always just one point of intersection exists. This is also true for the limiting cases of a perfect rigid wall—$\mathscr{S}'$ runs vertically, since $u_2 = 0$—and of a vacuum—$\mathscr{S}'$ runs horizontally, since $p_2 = p_0 = 0$—. In any case, either the hard or the soft reflection occurs, or the limiting intermediate case of complete transmission without any reflection. In this latter case, indeed, $p_1 = p_2$ and $u_1 = u_2$.

The distinction between *hard* and *soft reflection* is based on the difference of the slope of the two Rayleigh lines. We have for

$$\theta' = \varrho_0' S' \gtrless \theta = \varrho_0 D : \left.\begin{matrix}\text{hard}\\ \text{soft}\end{matrix}\right\} \text{reflection.}$$

The *shock impedances* θ and θ' are measures of the hardness of the two media. This is obviously a generalization of the acoustic impedance $\varrho_0 a_0$ and $\varrho_0' a_0'$ which are the limiting values for weak shocks. The above discussion is of course invariably applicable if $\mathscr{D}$ is an ordinary shock wave. θ', however, is not a material constant. It rather results only after the *u–p* construction has been performed. Therefore, we may use the terms *hard* or *soft material* only with some reservation.

The construction of the intersection in the u–p plane is applicable to many processes for which at a certain instant $t = 0$ two semispaces are in contact along a plane $x = 0$ without being in mechanical equilibrium; see the points 0′ and 1 in Fig. 24.2. Those processes could be, e.g., the head-on interaction of two shock waves, the impact of two solid bodies, the rupture of a membrane (in a shock tube) etc. In some cases $\mathcal{S}'$ has to be supplemented also by $\mathcal{E}'$.

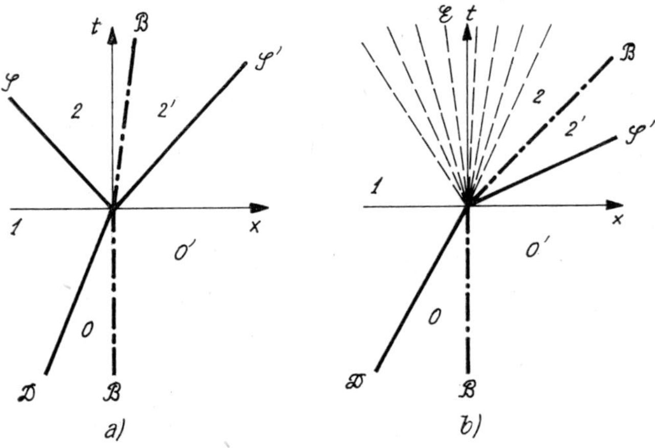

Fig. 24.2 Space-time plot (wave diagram) for head-on reflection; a) hard, b) soft.

The normal interaction of a detonation front with an inert material is technically used as a *measuring method* in order to determine experimentally the CJ values and the equation of state of the reaction products. For this purpose a special arrangement, a so-called *plane-wave generator*, is used; Fig. 24.5. With this arrangement, a limited and (except for boundary disturbances) plane, quasi steady CJ detonation front is generated. The shock velocity S' can be measured within the inert material by means of proper probes. If in addition the flow velocity u_2 behind the shock front is known p_2 is found as, compare (3.6),

$$p_2 = p_0 + \varrho_0' S' u_2 . \tag{24.2}$$

Here p_0 is negligible, and u_2 can be obtained from the free-surface velocity (see below). If, however, the Hugoniot curve of the inert material is known only measurement of S' is required in order to determine u_2 and p_2. In this way a point 2 is obtained in the u–p diagram, see Fig. 24.1c. Repetition of the experiment with a number of different inert materials of diverse hardness makes it possible to determine a fairly large part of the curve $\mathcal{S} + \mathcal{E}$ for the particular explosive; see Figs. 24.3 and 24.4. Intersection with the Rayleigh line which has the known slope $\theta = \varrho_0 D$ yields the CJ point 1. Transformation from u-p coordinates into v, p according to (3.5) and (3.6) gives the variation of the mechanical quantities along the isentrope downward from the CJ point and along the shock adiabat from CJ upward.

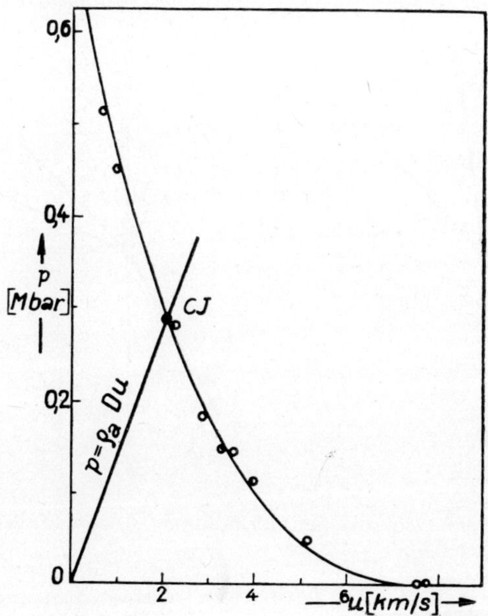

Fig. 24.3 Adiabat for the detonation products of composition B in a u-p diagram (from Ref. [70]).

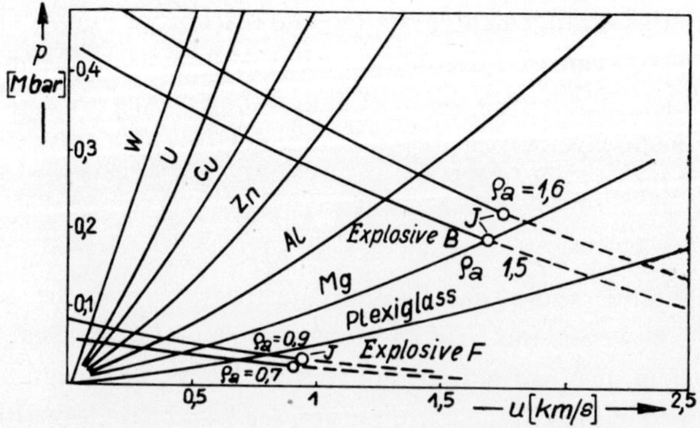

Fig. 24.4 Head-on interaction between various explosives and inert materials (from Ref. [53]).

In Fig. 24.4 the curves $\mathcal{S}'$ for a number of different inert materials of diverse hardness are shown together with curves $\mathcal{S} + \mathcal{E}$ of two different explosives, each for two different initial densities ϱ_0. The most compliant inert materials, namely the gases, cannot be represented in Fig. 24.4. The two lowest measuring points in Fig. 24.3 result from experiments with gases.

24.2 *The free-surface method*

For normal reflection of a strong shock at a boundary metal-vacuum or metal-atmosphere an expansion wave $\mathcal{E}$ is reflected which reduces the pressure in the metal to $p_2 = p_0$, see Fig. 24.1c. Within wide limits we may assume here that the shock adiabat of the metal deviates only slightly from the isentrope. In this case the u–p curves of the incident and the reflected wave ($\mathcal{D}$ and $\mathcal{E}$ in the diagram on the right-hand side of Fig. 24.1c) are symmetrical to the vertical line $u = u_1$ and it is $u_2 \approx 2u_1$. This behavior is fundamental for the application of the socalled *free-surface method*, see Fig. 24.5.

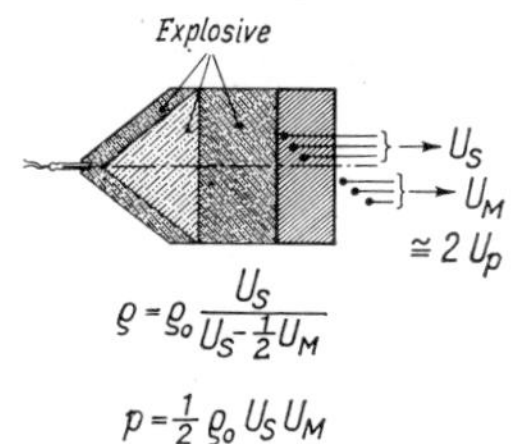

Fig. 24.5 The free surface method (from Ref. [106]).

Since the shock velocity U_s and the material velocity at the free surface U_M are directly measurable, not, however, the flow velocity or particle velocity U_p behind the shock, the pressure amplitude and the condensation are derived from U_s and U_M. So in (24.2) u_2 is replaced by $\frac{1}{2} u_M$.

In Fig. 24.5 in the lefthand part, a plane-wave generator is shown which is built out of two explosives with different detonation velocities (see Section 27).

For a detonation in a rarefied gas without casing, e.g. in the higher atmosphere, the escape velocity of the detonation products reaches practically the maximum velocity u_m which is obtained theoretically for expansion into a vacuum. This value is according to Fig. 24.3 $u_m \approx 8$ km/s. If we can assume calorically perfect gas a simplified approximate formula results for a very strong shock ($p_2 \gg p_0$) in a rarefied atmosphere. We obtain for this case

$$\frac{p_2}{p_0} = \gamma \frac{\gamma + 1}{2} \left(\frac{u_m}{a'_0} \right)^2$$

which results from formulas (3.6), (3.1), (3.19), (3.22). With $u_m = 8$ km/s, $a_0 = 320$ m/s, and $\gamma = 1.4$ we find e.g. $p_2/p_0 \approx 1050$.

With decreasing ambient pressure the shock front thickness increases according to the increase of mean free path of the molecules and the front becomes less sharp. For a detonation proceeding in a vacuum, that means $p'_0 = 0$, no shock occurs. The front of the explosion products where $p = 0$ propagates with u_m.

25 The Steady Oblique Detonation

25.1 *Coordinate transformation*

Suppose a plane quasisteady shock or detonation front $\mathscr{D}$ is incident on a plane boundary $\mathscr{B}$ with an angle of incidence $\alpha \neq 0$. In the *regular* case then, which we discuss at first, the incident wave, the refracted and reflected wave together with the undisturbed and the disturbed boundary form a steady and also pseudostationary configuration. We choose the coordinate system so that the z axis coincides with the intersection line of the primary wave and the boundary plane and the y axis is perpendicular to the undisturbed boundary, see Fig. 26.1. Figure 24.2 may be used again if we replace x by y and

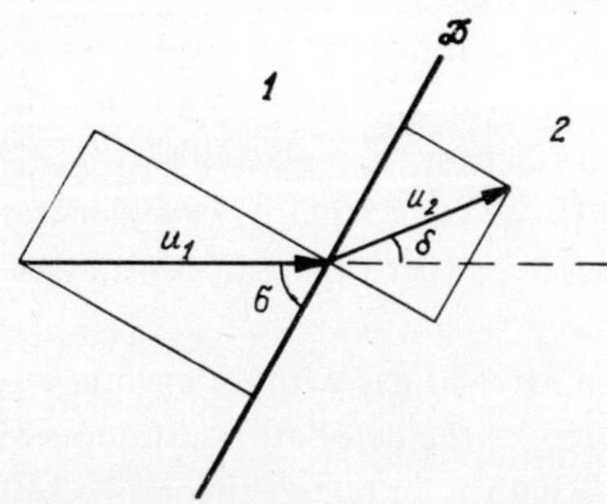

Fig. 25.1 Flow pattern for an oblique shock wave. $\mathscr{D}$ detonation or shock front; σ shock angle (of incidence); δ deflection angle.

t by x. Both media stream forward into the x direction with the velocity

$$u = \frac{D}{\sin \alpha}. \tag{25.1}$$

Henceforth we mark the upstream state by the index 1 instead of 0.

In order to describe the oblique reflection we list in the following at first the formulas for an oblique shock or detonation front in a steady flow. The appropriate description is derived from the fact that an oblique shock becomes a normal shock if observed from a coordinate system moving uniformly with the proper velocity along the shock front. Its flow pattern, see Figure 25.1, is therefore obtained from the former one-dimensional normal pattern, Fig. 1.1, by superimposing a constant velocity u_t tangential to the shock front. The components normal to the shock front remain unchanged. Hence

$$u_1 \cos \sigma = u_2 \cos (\sigma - \delta) \quad (= u_t). \tag{25.2}$$

The normal component of the upstream velocity is equal to the shock propagation velocity in the undisturbed medium

$$D = u_1 \sin \sigma. \tag{25.3}$$

This component forms the shock Mach number

$$M_S = D/a_1 \tag{25.4}$$

which is different from the incident flow Mach number $M_1 = u_1/a_1$,

$$M_S = M_1 \sin \sigma. \tag{25.5}$$

Correspondingly to (1.1) we have furthermore

$$\varrho_1 u_1 \sin \sigma = \varrho_2 u_2 \sin (\sigma - \delta) \quad (= \theta). \tag{25.6}$$

Resulting from (25.2) and (25.6)

$$\varrho_1 \tan \sigma = \varrho_2 \tan (\sigma - \delta), \tag{25.7}$$

or with the condensation ratio

$$\mu = \varrho_2/\varrho_1 = v_1/v_2 > 1 \tag{25.8}$$

$$\mu = \frac{\tan \sigma}{\tan (\sigma - \delta)},$$

$$\tan \delta = \frac{(\mu - 1) \tan \sigma}{\mu + \tan^2 \sigma}. \tag{25.9}$$

Since for each shock $D > a_1$, or $M_s > 1$, the shock angle is according to (25.5) always greater than the *Mach angle* of the incident flow:

$$\sigma > \arc\sin \frac{1}{M_1} = \alpha_{M1}. \tag{25.10}$$

For the (possibly overdriven) detonation moreover

$$D \geqq D_J, \quad \sigma \geqq \arc\sin \frac{D_J}{u_1} = \alpha_J, \tag{25.11}$$

where D_J denotes the CJ-detonation velocity and α_J a quantity defined analogously to the Mach angle.

25.2 *Polar diagrams for shock and reaction fronts*

Frequently for a certain upstream condition $(u_1, \varrho_1, p_1, \ldots)$ the entire family of possible steady oblique shocks has to be considered. This family of shocks is a continuous manifold with the parameter σ. Its range is according to (25.10), (25.11)

$$\alpha_{M1} \leqq \sigma \leqq \frac{\pi}{2}, \quad \text{(shock)} \tag{25.12}$$

$$\alpha_J \leqq \sigma \leqq \frac{\pi}{2} \quad \text{(detonation)}. \tag{25.13}$$

As a geometrical representation of this one-parameter family A. BUSEMANN has introduced 1931* the shock polar, Fig. 25.2.

This is a curve traced by the vector $\vec{u}_2$ of the downstream velocity (behind the shock) in the *hodograph plane* with the polar coordinates u and δ or the Cartesian coordinates

$$u_x = u \cos \delta, \quad u_y = u \sin \delta.$$

In the case of a calorically perfect gas the shock polar is an algebraic curve described by

$$u_{2y}^2 = \frac{(u_1 u_{2x} - a^{*2})(u_1 - u_{2x})^2}{a^{*2} + 2/(\gamma + 1)\, u_1^2 - u_1 u_{2x}}. \tag{25.14}$$

The *critical sound velocity* a^* is a constant which depends on the upstream characteristics. In a dimensionless representation the shape of the curve is

* See the references listed in [1].

determined only by γ and M_1. In gases with fixed γ, e.g. for all calorically perfect diatomic gases, for the determination of all oblique shocks only a single shock polar diagram with ∞^1 polar curves is required.

For *real gases* neither the analytic representation (24.14) nor the mentioned simplification is valid. For each gas there exist ∞^3 states for the approaching stream and just as many shock polars.

The shock polar is in general obtained as follows: For given p_1, v_1, u_1, and σ variable in the range (25.12), (25.13), θ is calculated according to (25.6). In the v_2-p_2 plane the Rayleigh line, see (3.8),

$$\frac{p_2 - p_1}{v_1 - v_2} = \theta^2$$

is brought to intersection with the Hugoniot curve determining hereby v_2 and p_2 as functions of σ. Furthermore M_s, μ, δ, u_2 are found from Eqs. (25.5), (25.8), (25.9), (25.6).

For $\sigma \to \alpha_{M_1}$ ($P \to A$ in Fig. 25.2) the shock degenerates into a Mach line; μ approaches one ($\mu \to 1$) and with (25.9) therefore $\delta \to 0$. For $\sigma = \pi/2$ ($P \to C$ in Fig. 25.2) δ becomes zero too, according to (25.9). Within the interval (25.12), however, we have of course $\delta > 0$. Consequently, there exists a point on the shock polar for which $\delta = \max$. In Fig. 25.2 obviously the *maximum deflection* is obtained when the line OP becomes tangential to the shock polar ($P \to T$). For given upstream conditions a deflection $\delta > \delta_{\max}$ cannot be performed by an oblique shock.

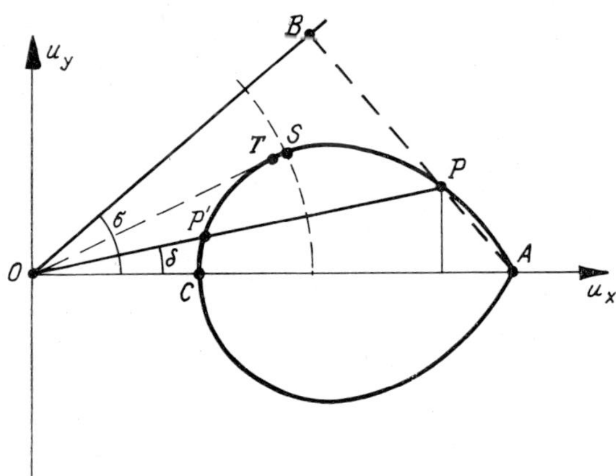

Fig. 25.2 Shock polar in the hodograph plane (Busemann 1931). $\overline{OA} = u_1$, $\overline{OP} = u_2$, $\overline{OB} = u_t$, $\overline{AB} = u_1 \sin \sigma$, $\overline{BP} = u_2 \sin (\sigma - \delta)$, $\overline{OC} \cdot \overline{OA} = a^{*2}$, $OS = a^*$.

For each $\delta < \delta_{max}$ there are two shock strengths which would lead to this deflection, corresponding to the two points P and P' in Fig. (25.2). P, together with the larger value u_2 and the smaller values of σ, μ, and p_2 characterizes the *weak solution*, while P' represents the *strong solution*.

The *detonation polar* starts at $\sigma = \alpha_J$ immediately with a finite deflection $\delta_J > 0$. The maximum deflection occurs in the inner of the interval (25.13). The quantities p_2, ϱ_2, u_2, M_2 vary in general monotonically with increasing σ. For the qualitative behavior of the flow it is important, whether the downstream flow is supersonic or subsonic, that means there is a distinct difference for $M_2 \gtrless 1$. We want to show that each shock or detonation polar passes the sonic limit $M_2 = 1$.

It is $M_2 = M_1 > 1$ for $\sigma = \alpha_{M_1}$. For $\sigma = \alpha_J < \pi/2$ the normal-flow-component Mach number of the motion behind the shock is owing to the CJ condition

$$M_2 \sin(\sigma - \delta) = 1,$$

and therefore also $M_2 > 1$. For $\sigma = \pi/2$, that means for the normal steady shock or overdriven detonation front, $\sigma - \delta = \pi/2$ and

$$M_2 = M_2 \sin(\sigma - \delta) < 1.$$

Therefore each real shock and detonation polar starts with supersonic flow in the off-stream direction and terminates at $\sigma = \pi/2$ with subsonic flow. In the limiting case $\alpha_{M_1} = \pi/2$ or $\alpha_J = \pi/2$, respectively, the curve degenerates into a point and M_2 becomes equal to one.

For a calorically perfect gas the maximum deflection has subsonic off-stream characteristics. That means, in Fig. 25.2 the point T lies in the inner of the sonic circle with the radius $\overline{OS} = a^*$ and the sonic limit point S on the shock polar with $M_2 = 1$ belongs to the weak solutions.

The qualitative results derived from the shock polars for calorically perfect gases are in general also valid for real non-detonating gases.

The classical shock polars in the *hodograph plane*, which we have discussed in some more detail owing to their evident features, are, however, not the proper graphical tool for our present purpose. With respect to the boundary conditions for p and δ which have to be satisfied in the case of oblique reflection (see Sec. 26) it is appropriate to transform the curves to rectangular coordinates p_2, δ or $\log p$, δ. The resulting curves are called "*Cardial curves*" after WEISE (1943)* or also shock polars [1].

* Reference cited in [1] (Cardial—translation of the German word "Herz-").

In the following we list some relations for oblique shock waves and detonations which are valid for arbitrary equations of state.

From Eq. (3.6) valid for a normal shock or from $p_2 - p_1 = \varrho_1 u_1^2 (1 - \mu^{-1})$ it follows for oblique shocks, that

$$p_2 - p_1 = \varrho_1 u_1^2 \sin^2 \sigma (1 - \mu^{-1}). \tag{25.15}$$

Herefrom by means of (25.9)

$$p_2 - p_1 = \frac{\varrho_1 u_1^2}{1 + \cot \sigma \cot \delta} \tag{25.16}$$

results. For given upstream conditions, therefore, the pressure increase can be obtained immediately from the angles σ and δ.

For the deflection angle δ, which according to (25.9) depends on μ and σ, an estimation can be given: It is always

$$\sin \delta \leqq \frac{\mu - 1}{\mu + 1}. \tag{25.17}$$

The equal-sign in (25.17) is valid just if $\tan \sigma = \sqrt{\mu}$. In this case it is also $\delta = 2\sigma - \pi/2$.

In order to obtain more explicit results for the oblique detonation we make use of the former (Sec. 9) introduced perfect-gas model. The approaching gas flow is determined by three dimensionless numbers: The adiabat exponent, the Mach number, and a heat parameter, e.g. after DAMKÖHLER:

$$\gamma, \; M_1, \; \bar{Q}. \tag{25.18}$$

We assume that $Q = c_p T_1 \bar{Q}$ is given in advance as the chemical energy per unit mass to be released in the detonation or reaction front.* We will associate with the three values (25.18) a family of oblique reaction fronts with σ as parameter based on a condition which is listed in the following. Namely, it is required that for the shock Mach number M_s according to (25.4), (25.5) and for the Damköhler parameter the equivalent inequalities (9.10), (9.11), hold which is known from the discussion regarding the properties of the

---

* For substances in a metastable equilibrium which are not necessarily detonable it is preferable to use the terms "reaction front" or "shock-induced combustion" rather than (overdriven) detonation.

normal reaction front. These inequalities write

$$\bar{Q} \leqq \frac{(M_S - M_S^{-1})^2}{2(\gamma + 1)} = \bar{Q}_{\max}(M_S, \gamma), \tag{25.19}$$

$$M_1 \sin \sigma \equiv M_S \geqq \sqrt{\frac{\gamma + 1}{2} \bar{Q} + 1} + \sqrt{\frac{\gamma + 1}{2} \bar{Q}}. \tag{25.20}$$

$\bar{Q}_{\max}$ is due to (25.19), with γ = const, a monotonically increasing function of M_s which for given M_1 reaches its highest value at $\sigma = \pi/2$. The restricting condition for the values (25.18) is therefore, like (9.10)

$$\bar{Q} \leqq \frac{(M_1 - M_1^{-1})^2}{2(\gamma + 1)} = \bar{Q}_{\max}(M_1, \gamma). \tag{25.21}$$

The righthand side of relation (25.20) is according to (10.1) the CJ-shock Mach number which belongs to the values γ and $\bar{Q}$ and which we, like before, denote by M_{SJ} in order to differentiate this Mach number from shock Mach number M_s of the general (overdriven) detonation. We write:

$$M_{SJ} = \sqrt{\frac{\gamma + 1}{2} \bar{Q} + 1} + \sqrt{\frac{\gamma + 1}{2} \bar{Q}}. \tag{25.22}$$

Instead of (25.21) we therefore may write also

$$M_1 \geqq M_{SJ}. \tag{25.20'}$$

Condition (25.20′) states simply that in a steady flow of a detonable substance a detonation front can be stationary if and only if the approaching stream velocity is so large that no CJ detonation can propagate upstream.

After (25.20′) is satisfied we may use (25.11) or

$$\sin \alpha_J = \frac{M_{SJ}}{M_1} \tag{25.23}$$

for the definition of a *Jouguet angle* α_J which by (25.13) determines the admissible range for σ.

25.3 *The two-γ model*

In reference [89] *reaction front polars* are discussed extensively reporting the appropriate formulas and diagrams. This discussion includes also subsonic reaction fronts like combustion fronts, see Figs. 7.2 and 7.4, allowing for different values γ_1 and γ_2 of the flow in front of and behind the reaction front.

Oblique reaction fronts in perfect gases with $\gamma_1 \neq \gamma_2$ are discussed also in Refs. [90] and [91]. But in [92], [93], [94] and [39] a constant value of γ is assumed.

With two different values γ_1 and γ_2 taken into account a better adjustment to the physical reality can be made. The formulas, however, become complicated. It seems now possible to combine the generality of the γ_1-γ_2 model with the simplicity of the γ model. Regarding this possibility we mention that for the final state behind the reaction front only p_1, v_1, H_1, $u_1 \sin\sigma$, and γ_2 are essential while γ_1 is used solely to define the quantities a_1, Q, M_s, and $\bar{Q}$ which determine the state of the approaching stream:

$$a_1^2 = \gamma_1 p_1 v_1, \quad M_S = \frac{u_1 \sin\sigma}{a_1}$$

$$H_1 = \frac{\gamma_1 p_1 v_1}{\gamma_1 - 1} + Q = \frac{\gamma_1 p_1 v_1}{\gamma_1 - 1}(1 + \bar{Q}), \tag{25.24}$$

compare the formulas (25.3) to (25.5) and (9.2) to (9.4). In addition to (25.24) we point out that behind the reaction front chemical equilibrium shall exist and therefore only *thermal* enthalpy can be present, namely

$$H_2 = \frac{\gamma_2 p_2 v_2}{\gamma_2 - 1}.$$

By this relation the *zero level* of the enthalpy is fixed.

Now it is convenient to use the method of Sec. 18; indeed (18.3′) holds rigorously with $\Gamma = \gamma_2$, $E^0 = 0$; therefore, applying (18.4), (18.6), (18.10) we define

$$Q = H_1 - \frac{\gamma_2 p_1 v_1}{\gamma_2 - 1}, \quad \bar{Q} = \frac{\gamma_2 - 1}{\gamma_2 p_1 v_1} Q \tag{25.25}$$

instead of (25.24) and

$$M'_S = \frac{u_1 \sin\sigma}{\sqrt{\gamma_2 p_1 v_1}} = M_S \sqrt{\frac{\gamma_1}{\gamma_2}} = M'_1 \sin\sigma \tag{25.26}$$

in addition to (25.23). By these stipulations we preserve our simple formulas derived earlier, namely the equilibrium-Hugoniot-curve equation (9.6) or (18.7) with $f = 2/(\gamma_2 - 1)$ and index 1 instead of 0; the inequalities (9.10), (9.11), (25.21) with γ_2 and M'_1; (25.19), (25.20) with γ_2, M'_1, M'_S, (25.20′) with M'_{SJ}; the CJ equations (10.1)–(10.4), (18.8), (25.22) with γ_2, p_1, v_1, M'_{SJ}.

The unreacted or frozen Hugoniot curve $\mathscr{H}_1 = \mathscr{H}_f$ can be calculated by (3.22) (with $\gamma = \gamma_1 = \gamma_f$, $p_0 = p_1$, $v_0 = v_1$). Since normally $\gamma_1 > \gamma_2$ there is an intersection of $\mathscr{H}_1$ and $\mathscr{H}_e$ like in Fig. 7.1.

The two-γ detonation model with $\bar{Q} > 0$ can be considered to some extent as a generalization of the two-γ relaxation model with $\bar{Q} = 0$, see Section 6.

For convenience the *heat parameter* F used in [39] may be introduced correspondingly to (9.19), (9.20) as

$$F = 1 + \sqrt{1 - \bar{Q}/\bar{Q}_{\max}(M_S', \gamma_2)}. \tag{25.27}$$

The case $F = 1$ corresponds to the CJ case and $F = 2$ to an equilibrium Hugoniot curve which runs through the initial point (p_1, v_1). According to (9.21) the pressure ratio for $1 \leq F \leq 2$ is given by

$$\frac{p_2}{p_1} = 1 + \frac{F\gamma_2}{\gamma_2 + 1}(M_S'^2 - 1). \tag{25.28}$$

Since, if written in a dimensionless form according to (25.15) with (25.26)

$$\frac{p_2/p_1 - 1}{1 - v_2/v_1} = \gamma_2 M_S'^2$$

we simply derive the condensation ratio μ from (25.28) as:

$$\mu^{-1} = \frac{v_2}{v_1} = 1 - F\frac{1 - M_S'^{-2}}{\gamma_2 + 1} = \frac{\gamma_2 + 1 - F + FM_S'^{-2}}{\gamma_2 + 1}$$
$$= \frac{(\gamma_2 - 1) + (2 - F) + FM_S'^{-2}}{\gamma_2 + 1}. \tag{25.29}$$

The values of μ and σ determine δ according to (25.9). Thus for fixed p_1, v_1, u_1, H_1, γ_2 and variable σ we obtain successively Q, $\bar{Q}$, M_1', M_S', $\bar{Q}_{\max}$, F, p_0/p_1, μ, δ and can accordingly draw the reaction front polar. The form of Eq. (25.29) shows under which conditions a very high condensation with $\mu \gg 1$ is obtained. This is the case exactly then, when simultaneously

$$\gamma_2 - 1 \ll 1, \quad 2 - F \ll 1, \quad M_S' \gg 1.$$

For the convenient formal application of the parameter F we indeed have to regard that a certain upstream condition is given uniquely by γ, M_1, $\bar{Q}$ not, however, by γ, M_1, F since $\bar{Q}_{\max}$ and F depend on σ. In general a physically reasonable shock polar can be attributed to each set of values of the three parameters γ, M_1, $\bar{Q}$ or γ_2, M_1', $\bar{Q}$, since we may assume that the chemical energy to be released is independent of the shock angle.

In Fig. 25.3 two typical *Cardial curves* are shown for shocks with and without reaction.

For the downstream Mach number we note, see [89],

$$M_2^2 = M_1'^2 \mu \frac{p_1}{p_2} (1 - (1 - \mu^{-2}) \sin^2 \sigma).$$

The letters

$$A\,(\sigma = \alpha_M), \quad C\left(\sigma = \frac{\pi}{2}\right), \quad T\,(\delta = \delta_{max}), \quad S\,(M_2 = 1)$$

have the same meaning in Fig. 25.3 as in Fig. 25.2; in addition J with $\sigma = \alpha_J$ occurs. S and T cannot be separated in the drawing.

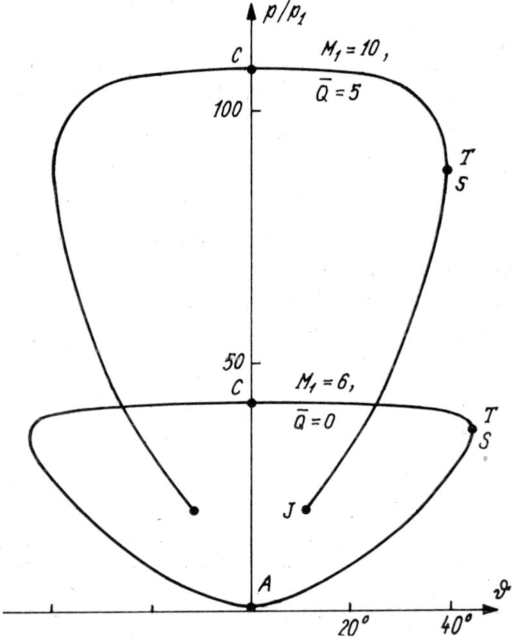

Fig. 25.3 δ–p polar ("cardial curve") for shock and detonation waves; $\gamma = 1.34$.

In some calculated examples it was always found that there was *one* point T with δ = maximum between J and C and that the points S and T were very close together.

26 Regular Oblique Reflection at a Plane Boundary

As pointed out at the beginning of Section 25 a close analogy exists between the normal and the regular oblique reflection. This analogy becomes very obvious if we compare Figs. 24.2 and 26.1.

The refracted wave $\mathscr{S}'$ is always a shock. The reflected wave is a shock $\mathscr{S}$ for hard reflection and a rarefaction wave $\mathscr{E}$ for soft reflection. This rarefaction wave or *steady centered expansion wave*, also called *expansion fan* or *Prandtl-Meyer flow*, is well known from classical gasdynamics.

Each of the three waves causes a pressure change and a deflection of the flow. Since at the boundary $\mathscr{B}$, as well before as after the reflection, the pressures on both sides must be equal, $p_0 = p_0'$, $p_2 = p_2'$, and also the flow directions, we obtain the necessary identities for the pressure ratios and the deflection angles:

$$\frac{p_{2'}}{p_{0'}} = \frac{p_1}{p_0} \cdot \frac{p_2}{p_1}, \quad \delta_{2'} = \delta_1 \mp \delta_2. \tag{26.1}$$

Counting all deflection angles positive the upper sign in (26.1) is valid for the hard reflection and the lower sign for the soft reflection.

The velocities u_2 and u_2' are in general different so that $\mathscr{B}$ downstream from the reflection point becomes a slip line or vortex sheet.

If the properties of the participating media and the direction and intensity of the incident shock $\mathscr{D}$ are given, the graphical construction using the p–θ plane results from (26.1); see Fig. 26.2 which is analogous to Fig. 24.1c.

The polar curves $\mathscr{D}$ and $\mathscr{S}'$ of the two media are fixed by the state of the approaching stream and particularly by the stream velocity (25.1). Actually

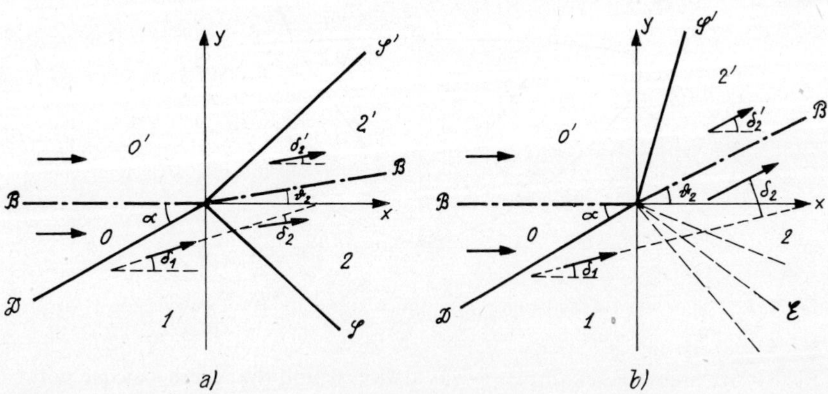

Fig. 26.1 Flow plane (cf. Figs. 24.1 and 24.2). a) hard, b) soft reflection. θ flow angle, δ deflektion angle. $\theta_1 = \delta_1$, $\theta_2 = \theta_2' = \delta_2'$, $\theta_0 = \theta_0' = 0$.

the state points 0 and 1 are already fixed and the polar $\mathscr{D}$ is not used in the construction furthermore. To state 1 as a new initial state corresponds a certain shock polar $\mathscr{S}$ which obviously in the θ–p plane has its foot at point 1. The intersection point 2 of $\mathscr{S}$ with $\mathscr{S}'$ (Fig. 26.2a) then determines the state 2, 2′ and the position and intensity of $\mathscr{S}$ and $\mathscr{S}'$ provided this point exists and is unique.

Similarly as in Section 24 the two cases of hard and soft reflection exist. In the latter case point 2 is the intersection point of $\mathscr{S}'$ with the rarefaction polar $\mathscr{E}$; see Fig. 26.2b.

At point 1 the tangents on $\mathscr{S}$, $\mathscr{E}$, and $\mathscr{D}$ have the same slope (except for the sign). This behavior is owing to the *contact relations* between the corresponding Hugoniot curves and the isentrope at point 1 of the v–p plane; compare Figs. 3.7 and 7.5.

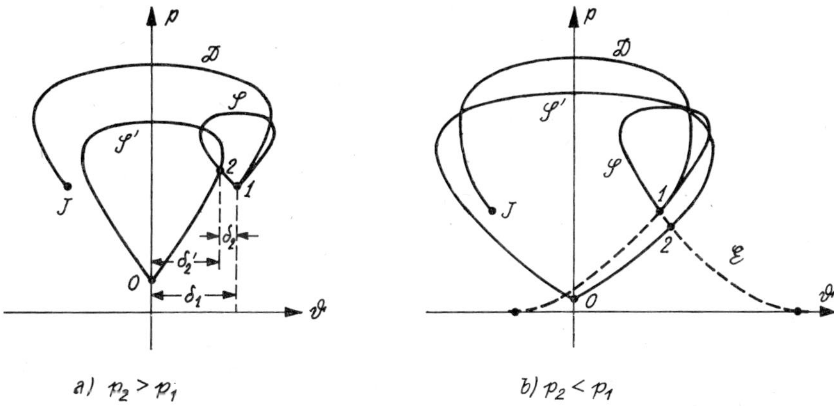

Fig. 26.2 Polar construction for Fig. 26.1. a) hard, b) soft reflection. The incident wave is assumed to be a CJ-detonation wave.

The polar $\mathscr{E}$ is monotonic terminating with a finite deflection angle at $p = 0$. We obtain the equation of the curve $\mathscr{E}$ by application of (25.9) to a very weak shock. We have then

$$\mu \approx 1, \quad \delta \approx 0, \quad \sigma \approx \alpha_M = \arcsin(1/M),$$

and we write

$$\delta \approx \frac{(\mu - 1)\tan\alpha_M}{1 + \tan^2\alpha_M} = (\mu - 1)\sin\alpha_M\cos\alpha_M,$$

$$\pm d\theta = \frac{d\varrho}{\varrho}\,\frac{\sqrt{M^2 - 1}}{M^2} = \frac{d\varrho}{\varrho}\,\frac{\sin 2\alpha_M}{2}. \tag{26.2}$$

Equation (26.2) is valid for infinitesimal condensation and rarefaction waves, for systems of simple waves in plane, isentropic, steady flow and particularly for the here occurring *Prandtl–Meyer expansion fans*. Together with the additional differential relations

$$dp = a^2 \, dp, \quad u \, du + \frac{dp}{\varrho} = 0 \tag{26.3}$$

which are valid along the streamlines in a steady isentropic flow and the appropriate equation of state, Eq. (26.2) can be integrated in principal. For calorically perfect gases where $a^2 = \gamma p/\varrho$ and with (26.3)

$$p\varrho^{-\gamma} = \text{const}, \quad u^2 + \frac{2}{\gamma - 1} a^2 = \text{const}.$$

Equation (26.2) can be integrated also in closed form as is well known; see e.g. [3], [5], [8].

In the case of normal reflection, where $\alpha = 0$ (see Fig. 24.1c), the existence and unique character of the intersection point of $\mathcal{S}'$ and $\mathcal{S} + \mathcal{E}$ was always certain; this is obviously not true for the *Cardial polar curves*; see Fig. 26.2. The regular solution discussed here corresponding to Fig. 26.1 is *not* existent in all cases for given states 0, 0′, the shock $\mathcal{D}$, and the angle α. The analogy between normal and oblique reflection is, of course, limited.

Regarding at first the unique character of the intersection point 2, Fig. 26.2, this uniqueness is obtained by the additional condition that only the lower monotonic section of the curves and $\mathcal{S}$ and $\mathcal{S}'$ is used; section A to T in Fig. 25.3. This, as we have seen, essentially results in the requirement that supersonic flow shall exist behind the shocks $\mathcal{S}$ and $\mathcal{S}'$. This property is important for the *stability* of the configuration according to Fig. 26.1. If in one of the downstream regions 2 or 2′ a Mach number $M < 1$ occurs, disturbances will propagate there streamupward changing the wave configuration, since the boundary conditions are not homogeneous up to infinity. At least, a shock bounding a subsonic region is in general curved.

The analogy between normal and oblique reflection is perfect in the limit $\alpha \to 0$. Here, owing to (25.1) $u_0 \to \infty$, $M_0 \to \infty$, $M_0' \to \infty$ holds. The flow is *hypersonic* and the well known *hypersonic similarity* can be applied. Evidently, the limiting process $\alpha \to 0$ must lead to the normal case $\alpha = 0$. For $\alpha \to 0$ all streamlines and wave fronts in Fig. 26.1 become parallel to the x axis so that this representation is no longer applicable. By means of the transformation

$$y = x', \quad x = \frac{D}{\sin\alpha}\,t'$$

to coordinates x', t' and succeeding limiting process $\alpha \to 0$ we are led back to Fig. 24.2 which in so far, together with the pertaining considerations, is applicable as approximation for the oblique reflection in the case $\alpha \ll 1$.

For finite α there exists still an extensive qualitative analogy between the normal and oblique reflection as long as for the oblique reflection $M > 1$ in all sections. The differential equations governing the one-dimensional unsteady flow are, of course, always of the *hyperbolic type*, the equations for the plane steady flow, however, are hyperbolic only in the supersonic regions. When $M = 1$ is reached the analogy breaks down completely and a regular reflection is no longer possible.

In the case of the regular hard reflection obviously two real intersection points of $\mathscr{S}$ and $\mathscr{S}'$ exist in general. Namely, besides the point 2 in Fig. 26.2a there exist another intersection point at higher pressure with subsonic flow in one of the off-stream regions. This ambiguity already occurs for the reflection at a rigid wall where the polar $\mathscr{S}'$ degenerates into a straight line, the p axis. In the earlier discussion we have called the second solution due to the stronger reflected shock the strong solution and the first with supersonic flow in the down-stream region the weak solution. Only the weak solution continuously approaches the unsteady normal reflection for $\alpha \to 0$.

The instability of the strong solution may be understood as follows. The boundary conditions which exist downstream of the regions 2 and 2′ (Fig. 26.1a) are in general not sufficient to maintain the higher pressure of the strong solution. In case the strong solution would be realized temporarily, rarefaction waves would propagate streamupward reaching the reflection point and decreasing the pressure. The wave configuration would collapse. The weak solution, on the contrary, is, due to the supersonic offstream conditions, not sensitive to such disturbances.

For the supersonic flow over a concave corner, see Fig. 26.3b, or over a wedge-shaped profile, see Fig. 26.3c, the same alternative occurs between the weak and strong solution as for the oblique shock reflection at a rigid wall, see Fig. 26.3a.

In all these three cases the upstream conditions are given with u_1, p_1, ϱ_1 together with the prescribed deflection angle δ_2. In the case $\delta_2 < \delta_{\max}$ a weak and a strong solution are obtained (points P and P' in Fig. 25.2) with the weak solution as dominant solution. For $\delta_2 > \delta_{\max}$ the solutions represented in Fig. 26.3 do not exist.

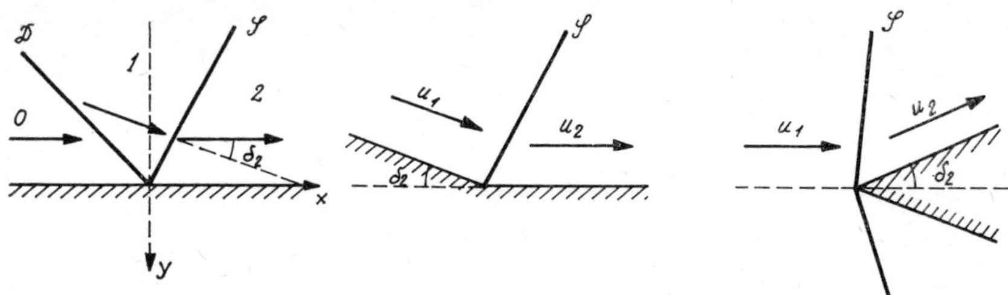

Fig. 26.3 The weak and strong solution: cases a), b) c).

Also in the case of soft reflection, see Figs. 26.1b and 26.2b, formally more solutions than one are possible. In general, the solution with the lowest pressure p_2 is then valid. If a soft solution and a hard solution appear to be possible, preference is to be given to the soft case.

Similarly as for the normal reflection also for the oblique reflection the limiting case of reflection-free *total absorption* exists as intermediate case between the hard and the soft reflection. For oblique reflection as compared with normal reflection it is even less possible to distinguish between *soft* and *hard* in terms of material constants since besides the shock intensity also the angle of incidence α is of influence.

The regular oblique reflection of a detonation front $\mathscr{D}$ at a material boundary $\mathscr{B}$ according to Fig. 26.1 is still existent if α reaches the value $\pi/2$, see Fig. 26.4.

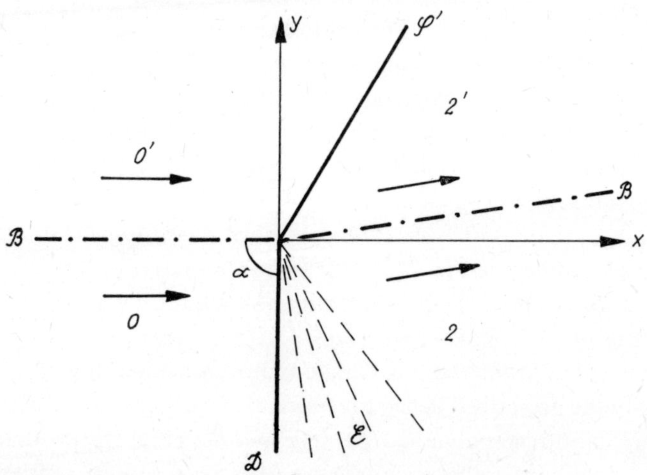

Fig. 26.4 Grazing interaction.

Behind $\mathscr{D}$ we then have $M_1 = 1$; section 1 in Fig. 26.1b disappears and the rarefaction $\mathscr{E}$ is directly attached to $\mathscr{D}$. For this *grazing* incidence with the wave normal parallel to the boundary surface always soft reflection occurs.

The configuration shown in Fig. 26.4 is always found for an elongated prismatic or cylindrical explosive charge which is cased into an inert material and detonates in the longitudinal direction, provided, only the immediate vicinity of the reflection point is considered, that means the curved lines are replaced by their tangents. In Fig. 26.5 an X-ray flash photograph of such a process is shown.

An explosive plate is covered on both sides with a layer of sand. The pour-density ϱ_0 of the sand is slightly different on both sides. We can clearly recognize the lines $\mathscr{B}$ and $\mathscr{S}'$. From the angles and the known detonation velocity the propagation velocities U_S of the shock waves in the sand result.

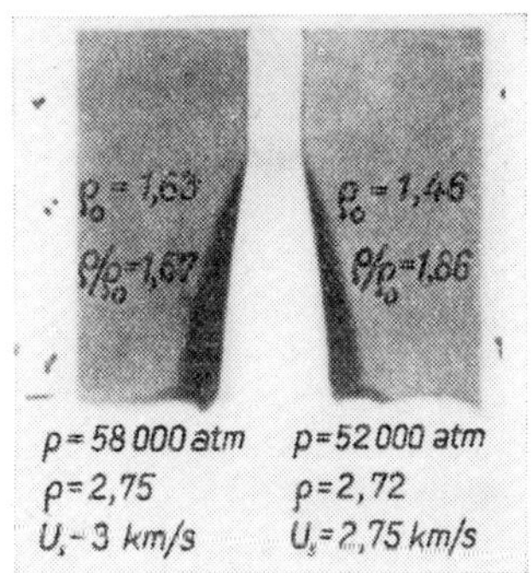

Fig. 26.5 Detonation between two sand layers (photograph ISL).

The condensation ϱ/ϱ_0 can be determined from the photographic density graduation. The pressure increase follows then according to (3.8) as

$$\Delta p = \varrho_0 U_S^2 \left(1 - \frac{\varrho_0}{\varrho}\right),$$

without requiring thermodynamic state relations of the inert material or the detonation products. Since the X-ray absorption of the detonation products is very small the tail of the expansion wave $\mathscr{E}$ cannot clearly be seen.

Figure 26.6 shows an experiment using a hard-rubber plate as inert material with an explosive layer attached on both sides. While the explosive and the detonation products are unvisible in this photograph and the position of the detonation front can be localized only indirectly (see the sketch) the strong slightly curved shock waves in the hard rubber can clearly be seen as well as their *regular intersection*, see also Section 27.

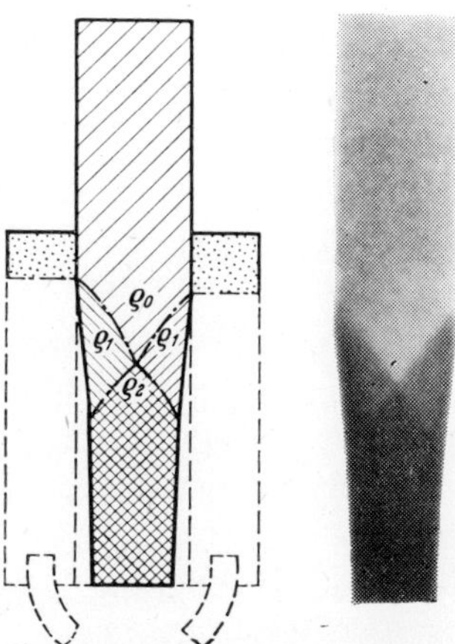

Fig. 26.6 Hard-rubber plate with detonations on both sides (photo ISL). Sketch and X-ray photograph.

27 Further Regular Interactions

So far we have made the assumption that a detonation front strikes the boundary of a second inert material and that also the first medium after the detonation has passed (that means the detonation products) behaves like an inert material. Besides this configuration many other related processes which can be treated similarly are conceivable, and several of those are practically important. We will call a wave configuration a *regular interaction* if the x–t plane as in Fig. 24.2 or the x–y plane as in Fig. 26.1 appears subdivided into sectors by detonation fronts, shock waves, expansion waves, and material boundaries in the sense of a pseudostationary configuration. Herein, in order to simplify and following [91], we describe material boundaries temporarily also as waves.

For example, also the second medium may be detonable but with different properties, or a moderately intense shock propagating in the first medium may be reflected at a hard boundary and return into the first medium as a detonation front with the second medium being an explosive or an inert material, or a detonation front may collide with a shock, or an interference of two shocks may occur; cf. Fig. 26.6.

Reference [91] gives a systematic survey of the possible cases of a regular oblique interaction. Thereby it is assumed as a rule that two given incident waves collide at the reflection point and that two pressure waves as well as an intermediate limiting line (slip line) originate from this point extending downstream. There exist, however, certain exemptions from this rule. For the quantitative description always intersections of θ–p polars are used. A corresponding discussion of the normal interaction seems not to exist. Without doubt, such a discussion would be quite analogous and in a less degree subject to restricting conditions. The θ–p curves would be replaced by u–p curves; cf. Fig. 24.1c. We do not repeat this systematic survey here. But an important and essential supplement is added to our discussion resulting from [91].

Suppose the second medium is detonable and the first an inert very rigid material so that for this medium always $u = 0$. This is a meaningful limiting case of a real material behavior. At $t = 0$ a shock wave from the first medium strikes the second medium head-on initiating at $x = 0$, $t = 0$ a detonation front which propagates into the x-direction. This is a CJ detonation since no energy is provided from outside behind the wave. In Sec. 12.3 (see Fig. 12.6) we have already indicated that the boundary condition $u = 0$ at the rigid wall requires a rarefaction.

What actually happens is known to us from the discussion in Section 23. We have here the case which was characterized there by $\nu = 1$ with the boundary condition (23.12). The solution found there was a CJ detonation with an immediately following rarefaction wave. This configuration was capable to satisfy the boundary conditions there, either (23.12) corresponding to a rigid wall, or (23.24) corresponding to vacuum with $p = 0$.

The CJ detonation with succeeding rarefaction of arbitrary constant intensity has to be considered as an entity, so to say as a kind of *weak detonation* which occurs besides the strong, overdriven detonation. This detonation has a definite final state but a zone length increasing with time.

In a v–p diagram, cf. Fig. 7.5, the final states behind a CJ detonation with expansion (along the isentrope $\mathscr{I}$ downward from CJ) match without bend those behind a strong detonation (on $\mathscr{H}_e$ from CJ upward). Thereby a *pseudo Hugoniot curve* $\mathscr{H}_e + \mathscr{I}$ is composed which appears as a shock polar $\mathscr{D}' + \mathscr{E}'$ without bend in the u–p diagram and the δ–p diagram, cf. Figs. 27.1, 27.2.

A shock wave propagating in the first inert material and incident head-on on the surface of the second, explosive material produces in general—provided initiation occurs—either an overdriven detonation or a CJ detonation with succeeding rarefaction. The reflected disturbance can be a shock wave or a rarefaction wave.

Therefore, four cases result which are distinct from each other in the u–p diagram, cf. Fig. 27.1, according to whether $\mathscr{S}$ or $\mathscr{E}$ form an intersection point with $\mathscr{D}'$ or with $\mathscr{E}'$. The reflection-free case, between $\mathscr{S}$ and $\mathscr{E}$, as well as the pure CJ detonation, between $\mathscr{D}'$ and $\mathscr{E}'$, are limiting and exceptional cases. Since the curves $\mathscr{S} + \mathscr{E}$ and $\mathscr{D}' + \mathscr{E}'$ in the u–p diagram are running monotonically from $p = 0$ up to $p = \infty$ the existence and uniqueness of the

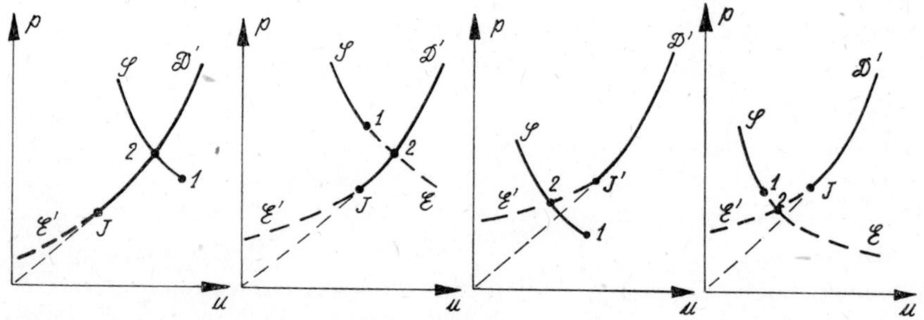

Fig. 27.1 Initiation of a detonation through an incident shock wave.

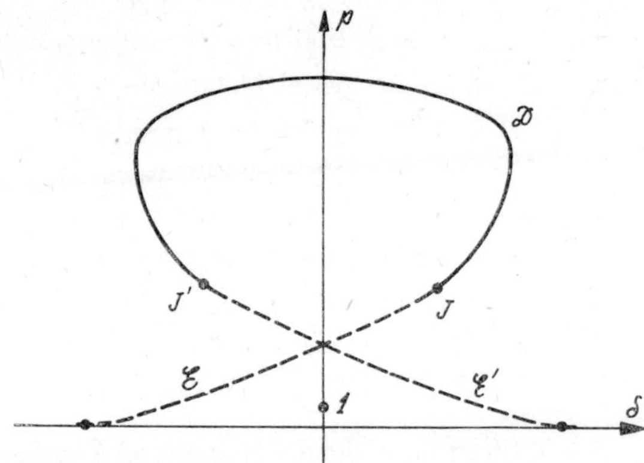

Fig. 27.2 Complete detonation polar.

intersection point is guaranteed. The impedance construction with detonations becomes formally identical with the corresponding construction for ordinary shocks and expansions. Only a CJ front with succeeding rarefaction is to be understood as a *single* wave. The rule stated above that after the reflection only one wave propagates in each of the media has to be interpreted in this sense.

If the first "medium" is a vacuum, then $\mathscr{S} + \mathscr{E}$ in Fig. 27.1 is replaced by the u axis. The expansion $\mathscr{E}'$ extends from J down to $p = 0$. This is the case of the plane Taylor wave with the boundary conditions (23.24), (23.25).

As an example of the regular wave interaction we also have to add the head-on collision of two condensed materials. At the instant $t = 0$ we have, similarly as for an incident shock, see Fig. 24.2, two media in contact at $x = 0$ which have the states 0′ and 1 respectively. In this case, e.g., we may have $p_0' = p_1 = 0$, $u_0' = 0$ but $u_1 > 0$. The points 0′ and 1 are at the same level in the u-p diagram. The construction of the intersection points yields clearly two shocks $\mathscr{S}$ and $\mathscr{S}'$. Each of those can be in principle a detonation if the medium is detonable. The *initiation* of a detonation by means of the head-on collision of a metal plate and an explosive is frequently used as an experimental tool. In the first instant the detonation is either a strong detonation or a CJ detonation with succeeding rarefaction wave. Since for experimental conditions the propellent plate has always a small thickness the forced detonation is very soon caught by rarefaction waves and reduced to a CJ detonation.

We will not discuss here in detail the regular oblique interaction; reference is made to [91]. These cases are analogous to the normal interaction as pointed out before. We give, however, a sketch of the δ–p polar for the oblique detonation with possibly succeeding rarefaction wave; cf. Fig. 27.2.

Since the actual detonation polar $\mathscr{D}$ is a symmetrical curve with two CJ points J and J' we have accordingly also two corresponding expansion polars $\mathscr{E}$ and $\mathscr{E}'$. A certain initial state point 1 belongs to $\mathscr{D}$ similarly as to the Hugoniot curve from which $\mathscr{D}$ has been derived, while the curve $\mathscr{E}$ or $\mathscr{E}'$ alone may be followed downward originating from any of its points.

For the interaction of a shock wave (or a colliding body) with a detonable substance the intensity and duration of the impact determines whether a simple shock wave or a detonation wave is produced in the explosive. The decision between these *two possibilities* cannot be obtained by means of our polar diagrams and actually not at all within the scope of the theory of an ideal detonation; see Sec. 30.2.

The reflection and refraction of a detonation front at the boundary of two gases can be realized experimentally. Quantitative observations can be made if e.g. an explosive and an inert gas are separated by a thin membrane with negligible mechanical strength; cf. [95]. While for this experiment the produced propagating detonation wave would appear as steady only in a comoving frame of reference, an oblique detonation front standing in the laboratory system can be obtained in the steady supersonic flow of an explosive gas past a wedge; see [96]. The upstream velocity, however, must be of course higher than the CJ-detonation velocity, see (25.20′).

We briefly mention an important application of the transmission of a detonation into a second detonating substance. Since the detonation wave front has a characteristic constant propagation velocity in each homogeneous medium, as have light and sound waves, the Snellius law of refraction is valid for detonations. By means of two explosives of different CJ-detonation velocities, separated by a surface of an appropriate shape, a detonation front of a desired form can be obtained. The employment of such "explosive lenses" is sometimes called explosive or detonation optics [107].

28 Irregular Reflection

It has been pointed out before that for given initial and boundary conditions, in contrary to the case of normal interaction, for the *oblique interaction* a *regular solution* of the type discussed above does not always exist. We may illustrate the situation as follows.

Suppose a wave $\mathscr{D}$ approaches head-on (with zero angle of incidence) a boundary $\mathscr{B}$ (Fig. 28.1a). The region between $\mathscr{D}$ and $\mathscr{B}$ then remains undisturbed until the exact instant of collision. If, however, $\mathscr{D}$ is inclined by an angle α against $\mathscr{B}$ then it is not immediately obvious that the angular region

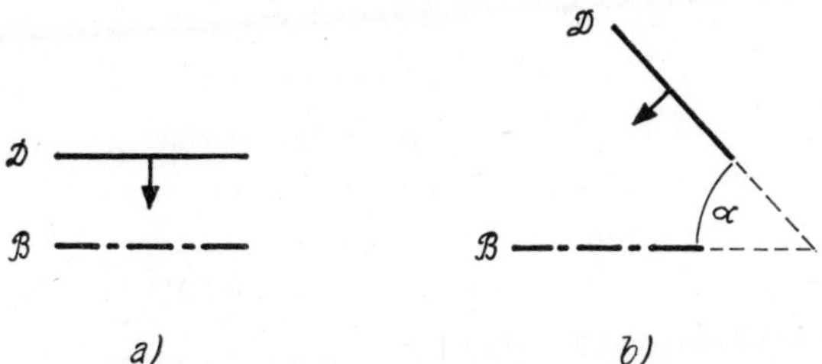

Fig. 28.1 Head-on and oblique reflection.

between $\mathscr{D}$ and $\mathscr{B}$, cf. Fig. 28.1b, is undisturbed exactly up to the intersection line of the two planes. It may be that part of this region is influenced by the interaction between $\mathscr{D}$ and $\mathscr{B}$ which has already begun. If this happens the regular wave configuration according to Fig. 26.1 is no longer possible.

Here at the beginning we point out that the irregular reflection is not a specific detonation problem. It rather occurs as well for ordinary shock waves. We therefore may restrict our discussion making reference to the available literature.

Starting from the now familiar range of the regular reflection we change the parameters, particularly α, so that we reach the limit beyond which the regular reflection ceases to exist. What happens at this point?

The regular solution is found according to Fig. 26.2 as the intersection point (2) of two polar curves $\mathscr{P}$, $\mathscr{P}'$ in the θ–p diagram; see Fig. 28.2. These curves start countercurrently from their initial points 1,0 extending as far as we are interested here from the θ axis up to the points T, T' having there a vertical tangent. In Fig. 28.2 we do no longer differentiate between shock and expansion polar. The curve sections which physically cannot be realized, however, have been rendered by dashed lines.

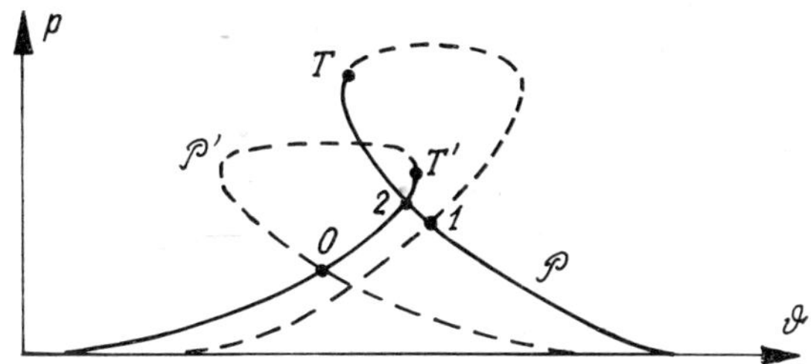

Fig. 28.2 The limit of the regular reflection.

Changing Fig. 28.2 continuously, the regular intersection point 2 obviously ceases to exist if it becomes identically with T or T'. Therefore, we have to distinguish two cases, depending on, whether the maximal deflection is reached in the first medium with 2 approaching T, or in the second medium with 2 approaching T'. Fig. 28.2 indicates the second case.

Mathematically, in fact, the intersection point 2 can still be constructed continuously until $\mathscr{P}$ and $\mathscr{P}'$ will merely contact each other. The contact point is then slightly above T or T'. The contact configuration, by the way, would deviate only slightly from the somewhat arbitrarily defined limiting configuration $2 = T$ or $2 = T'$. Equivalently we could let terminate the purely regular solution already for $2 \to S$ or $2 \to S'$ respectively with S, S' being the sonic limit points, cf. Fig. 25.2. These points lie as far as known on $\mathscr{P}$ and $\mathscr{P}'$ immediately below T and T', respectively.

The case $2 \to T$ or $2 \to T'$ occurs particularly then when M_1 or M_0' decrease approaching the value one and thereby the upper part $\mathscr{S}$ of $\mathscr{P}$, or $\mathscr{S}'$ of $\mathscr{P}'$, gradually reduces to a single point. For $M_1 = 1$, of course, no shock $\mathscr{S}$ exists and only the expansion polar $\mathscr{E}$ is left. If M_1 becomes smaller than one, also the expansion polar disappears. Behind a grazing incident ordinary shock*, however, that means for $\alpha \to \pi/2$ in Fig. 26.1, there is always $M_1 < 1$. It is possible also for each α that $M_0' < 1$ and therefore $\mathscr{P}'$ does not exist. These cases are already beyond the limit of the regular reflection.

We consider the *degenerative cases* $2 \to T$ and $2 \to T'$ more closely by selecting the cases with the most simple conditions. The associated phenomena are commonly called *Mach effect* and *total reflection*, cf. e.g. [97].

The first case is a phenomenon which belongs essentially to the first medium. In principle, it can be observed for the reflection at a rigid wall, cf. Fig. 26.3a. What happens if for the given upstream conditions (state 1) the required deflection δ_2 cannot be realized by means of a shock? In all cases *a*, *b*, *c* of Fig. 26.3 the same happens—best known for case *c*—: The shock moves forward, *separates* from the wedge and takes eventually a new stable position depending on the downstream conditions. Hereby $M < 1$ near the profile tip. In the case of Fig. 26.2a the forward motion of $\mathscr{S}$, while $\mathscr{D}$ keeps its position, leads to a *triple shock formation*, the so-called *Mach effect* or *Mach reflection*; cf. Fig. 28.3.

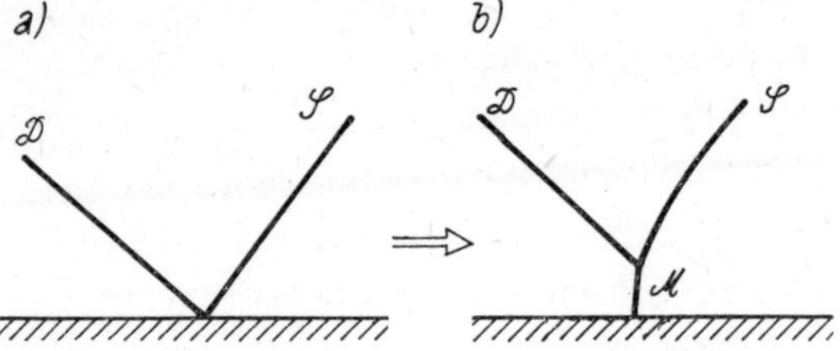

Fig. 28.3 The Mach effect at a rigid wall. (a) regular, (b) irregular (Mach) reflection. $\mathscr{D}$ incident shock or detonation wave, $\mathscr{S}$ reflected shock, $\mathscr{M}$ Mach shock (Mach stem).

If $\mathscr{D}$ is a shock wave with the pressure ratio p_1/p_0 propagating in a perfect gas then the limiting or *critical angle* α_c of the regular reflection can be given in analytical form as a function of p_1/p_0 and γ, [1]. At a rigid wall regular reflection occurs for $\alpha < \alpha_c$.

Behind the Mach shock $\mathscr{M}$ and the reflected shock $\mathscr{S}$, see Fig. 28.3b, the Mach number is $M < 1$. $\mathscr{M}$ and $\mathscr{S}$ are therefore curved. If $\mathscr{D}$ is a detonation then $\mathscr{M}$ represents a strong detonation with higher pressure than for $\mathscr{D}$.

---

* Different for a CJ detonation; see Fig. 26.4.

In case a Mach effect developes from a regular reflection at the boundary of a second real medium, this Mach effect in the first medium influences only little the refracted shock wave $\mathscr{S}'$ in the second medium; see Fig. 28.4.

$\mathscr{M}$ propagates along the boundary surface $\mathscr{B}$ drawing with it a shock $\mathscr{S}'$ in the second medium. The part of $\mathscr{B}$ extending downstream is curved. The same is true for $\mathscr{S}'$ although still $M > 1$ behind $\mathscr{S}'$.

The shock waves $\mathscr{M}$ and $\mathscr{S}'$ must satisfy three conditions at their common intersection point with $\mathscr{B}$. Namely, the deflection, the pressure increase and the intersection trace velocity must be equal. This would not be possible if

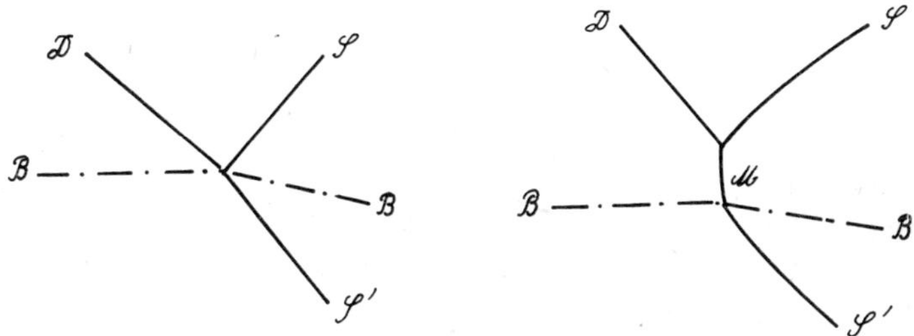

Fig. 28.4 Regular and Mach reflection with refraction. $\mathscr{S}'$ refracted shock.

the intensity and the angle of incidence of $\mathscr{M}$ would be prescribed. Suppose $\mathscr{M}$ is a shock wave of certain intensity, then caused by the subsonic region behind $\mathscr{M}$ the incidence angle of $\mathscr{M}$ on $\mathscr{B}$ is adjusted so that the three conditions are satisfied.

If $\mathscr{M}$ is a strong detonation in general it is reduced to a CJ detonation in the course of the process. Then, however, we will find $M \geq 1$ behind $\mathscr{M}$ and regular reflection occurs again; cf. Fig. 26.4.

In the second possible degenerative case, with $2 \to T'$, the refracted shock is "overstressed" by the intersection condition of Fig. 28.2 and is forced to separate from the reflection point and to move forward. After the shock has moved away a steady state can be established again with an incident and reflected wave remaining in the first medium while in the second medium, owing to $M < 1$, no refracted (steady) shock can exist. Strictly speaking, neither in the first nor in the second medium an undisturbed region exist.

Essential properties of this *irregularity* can be studied already with infinitesimally weak, *acoustic* plane pressure waves if the second medium has higher sound velocity than the first and if the angle of incidence exceeds the *critical angle* of the *total reflection*; cf. [97]. The acoustical total reflection which is strictly analogous to the corresponding optical casc can be described in a linearized form if we restrict the discussion to continous waves. Already arbitrarily *weak shocks*, however, cannot be treated in a linearized approach, they rather require new considerations. These circumstances are discussed in more detail in Ref. [97] and will not be further discussed here since these phenomena were so far of no particular importance for detonation processes.

The Mach reflection has been investigated theoretically and experimentally in numerous papers, mostly without relation to detonation phenomena or to a real second medium, and mostly for plane pseudostationary processes. Reference [98] gives a comprehensive review and Ref. [3] takes also detonations into account.

A more recent, predominantly experimental investigation is concerned with the interaction of detonation waves and strong shock waves within the solid explosive and the reaction products [99]. From the remaining deformation of a metal surface which was attached to the explosive surface before the detonation took place, the wave propagation can be reconstructed since the trace of each *triple-point* marks off as a sharp line on the metal. Evaluating the curves and angles the pressure values can be obtained and certain conclusions can be made regarding the equation of state of the detonation products.

The Mach effect is observed mainly in the case of hard reflection; it is, however, not restricted to this case. With increasing angle of incidence α the regular weak reflection ceases to exist exactly if behind the incident shock $M < 1$ occurs. A reflected steady rarefaction wave is then no longer possible. Unsteady rarefaction disturbances, however, are now propagating streamupward in the subsonic flow region resulting in an attenuation and curvature of the incident shock. The appropriate boundary conditions are satisfied at the point of incidence. The limiting point between the disturbed and undisturbed parts of the shock corresponds to the triple point of the hard Mach reflection.

An *ordinary detonation front* behaves differently in this respect as compared with a shock wave. The detonation discontinuity can be intensified by succeeding shock waves but it can not be attenuated by rarefaction waves. The region penetrated by rarefaction waves may extend directly up to the detonation wave, still the CJ pressure remains unchanged at the front; compare e.g. the *Taylor wave*, Fig. 23.3. These statements are of course valid only provided an *ideal detonation* can be assumed. This assumption, however, becomes questionable just where intense rarefaction waves occur (Ch. IX).

We have essentially restricted our discussion here to processes which are related directly to a detonation front incident on the boundary of an adjacent medium. The theoretical analysis of an entire blast process requires a considerable computation effort and our present discussion may lead at most to the proper initial conditions for this work.

The volume of the calculations depends on the geometrical configuration and on the mechanical-thermodynamic properties of the participating materials. Older investigations [100], [88], [101] suppose necessarily very simple geometric and thermodynamic conditions. Through advanced computer techniques later also more complicated configurations and more realistic material properties could be simulated by calculus. Some of these more recent contributions can be found in the conference proceedings [102], [103], [104], [105].

CHAPTER IX

The Non-Ideal Detonation

29 The Steady Non-Ideal Detonation

29.1 *The diameter effect*

Regarding the structure of the detonation wave we have assumed so far that the wave front is plane and the flow observed in the appropriate coordinate system is one-dimensional and steady. The additional requirement that the detonation wave should propagate self-sustained has led to the CJ condition. We now remove the first of these idealizing assumptions, namely, the planar character of the wave front. Within this special meaning we will talk of a *non-ideal detonation* as is commonly done. It turns out that also the second condition then can no longer be readily satisfied (see below).

A selfsustained detonation propagating along an extended cylindrical or prismatic charge can be described as steady in a comoving frame of reference.

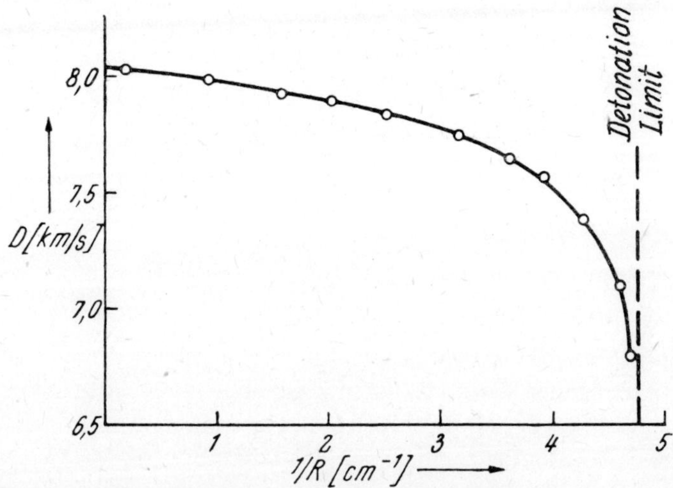

Fig. 29.1 Diameter effect for TNT/RDX 35/65 (from Ref. [107]).

There exist, however, some pecularities due to nonideality. We will describe those at first for the case of a circular-cylindrical charge without casing.

The detonation velocity D depends on the radius R of the cylindrical charge. The function $D(R)$ is a monotonically growing function. The limiting value

$$D(\infty) = \lim_{R\to\infty} D(R) = D_{CJ} \tag{29.1}$$

is the *ideal*- or CJ-detonation velocity.

There exists a limiting radius R_c where steady detonation is possible for $R > R_c$ but impossible for $R < R_c$. This behavior is termed the *diameter effect*. Plotting D versus $1/R$ we find an almost linear dependence for $R \to \infty$; cf. Fig. 29.1.

Cylindrical and prismatic charges with casing behave similarly. The functional relationships, however, are more complicated due to the influence of the density, thickness, and strength of the casing material.

A first qualitative explanation of the diameter effect results considering the steady flow pattern schematically shown in Fig. 29.2.

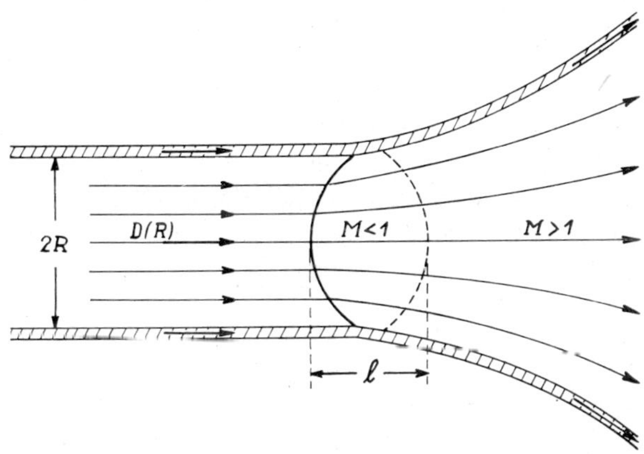

Fig. 29.2 Nonideal steady detonation of an explosive with casing.

The *shock front* has a convex curvature in the forward direction. The streamlines are bended sharply at the shock front and diverge in the reaction region. The reaction does not proceed with constant cross-sectional area as in the ideal case but rather in an expanding flow.

The shock angle along the shock front decreases from the inner to the outer parts and therefore the shock Mach number and the shock pressure also decrease. In other words, the rarefaction waves penetrating the reaction region from the outside attenuate the front shock from the boundary causing a shock curvature.

The detonation products are streaming in a quasirigid *nozzle* formed by the casing material. The flow which is initially subsonic behind the shock finally becomes highly supersonic at considerably decreased pressure. The downstreaming detonation products must therefore pass a *sonic limit.* The exact position of this sonic line is not known. It must, however, be within the reaction region as we shall see below. Without reaction this sonic limit would be, of course, at the smallest cross-sectional area. This, however, is not possible in the present case. According to [107] the subsonic region is probably lens-shaped, see Fig. 29.2.

Reactions which still proceed beyond the sonic limit have no mechanical influence on the shock front; the reaction energy which is released in the supersonic flow region does not contribute to the propulsion or acceleration of the shock. It is quite obvious that even at the center the front shock is weaker and propagates slower compared with the ideal case since only a reduced reaction energy is available.

The process is essentially *two-dimensional.* Particles close to the boundary go through a weaker shock and a faster expansion than those close to the axis and are therefore chemically transmuted less completely in the subsonic region. This applies particularly for charges without casing in a vacuum with the outer parts being kept constantly at low pressure.

29.2 *Semiempirical theories*

The theoretical description of the entire detonation process is very difficult for several reasons and could not be performed satisfactorily in any case. Only if one assumes considerably simplified *approximate methods* one may derive formulas describing relations which can be proved experimentally. Reference [108] is a classical example for such methods.

As a characteristic quantity for the non-ideal behavior of an explosive the *reaction-zone length* l* is introduced in [108]. This quantity is to be added to the quantities listed in Table 16.1. It cannot be derived from these other quantities. Even for an ideal CJ detonation l cannot be defined without arbitrariness since the reactions approach the equilibrium *asymptotically* and not after a finite length and time. The discussion in [108] which essentially takes the shock-front curvature into account and also determines the shape

* In [108] denoted by a.

of the shock front leads to a simple semiempirical formula

$$\frac{D(R)}{D_{CJ}} = 1 - \frac{1}{2}\frac{l}{R}, \tag{29.2}$$

which would be valid for sufficiently small values of l/R.

According to Fig. 29.1 Eq. (29.2) is indeed valid asymptotically for $R \to \infty$ with a constant l which is just defined by (29.2) representing at least a useful measure for the reaction zone length.

Resulting from [108] the experimental curves $D(R)$ yield according to (29.2), e.g., the values $l = 2.2$ mm for picric acid ($\varrho = 0.9$ g/cm^3) and $l = 0.826$ mm for *RDX* (0.9 g/cm^3).

For *cased* cylindrical charges Ref. [108] gives the relation

$$\frac{D(R)}{D_{CJ}} = 1 - 2.17\frac{(l/R)^2}{W_c/W_e}. \tag{29.3}$$

It is hereby assumed that the casing material has a damming effect through its inert mass only. Correspondingly the ratio W_c/W_e of the casing mass and the explosive mass enters the calculation.

In the case of a very *thick casing* which is not accelerated as an entity but rather predominantly compressed by the shock wave, relation (29.3) is no longer valid. For this case Ref. [108] gives the formula

$$\frac{D}{D_{CJ}} = 1 - 0.88\frac{l}{R}\cos\psi, \tag{29.4}$$

where ψ is the angle of incidence of the shock front interfering with the cylindrical boundary separating the charge from the casing material. Equations (29.3) and (29.4) yield for Amatol 60/40 ($\varrho = 1.55$ g/cm^3) $l = 3.98$ mm, for Minol 2 in lead $l = 5.38$ mm, for TNT ($\varrho = 1.6$ g/cm^3) $l = 0.36$ mm, and for TNT ($\varrho = 1.615$ g/cm^3) in steel $l = 0.346$ mm.

It is already suggested in [108] that the values for the reaction zone lengths resulting from (29.2)–(29.4) are too high. An independent absolute measurement of this quantity is, however, very difficult. The numbers determined by the method mentioned above are at least quite useful as relative values.

The reaction zone length l is not a material constant. Its value grows with increasing deviation from the ideal detonation behavior. Accordingly, $D(R)$ falls off with decreasing R more than linearly relating to $1/R$, cf. Fig. 29.1, as it would be the case due to (29.2) for constant l.

While in [108] the available experimental and theoretical information for the shock front curvature is essentially used, the somewhat older theory in [109], which is reviewed in detail in [108], does not need this information. The flow of the detonation products is divided schematically there into a central *nozzle flow*, treated by streamline approximation, and a *Prandtl–Meyer expansion* of the regions adjacent to the boundary. Different formulas result which also use the reaction zone length. Both theories are based on simplifications and the experimental results seem not to indicate a clear preference for one or the other of the two methods. The theory of [108] has been taken up and further developed in Ref. [53].

29.3 *The generalized CJ condition*

In the case of nonideal detonations the term *CJ plane* is also frequently used identifying this plane with the end of the reaction zone. This requires the following comments:

1) The true *CJ condition* is characteristic for the *ideal* detonation, cf. Chapt. V, and is not satisfied for a non-ideal detonation. For the latter the CJ state is nowhere attained.

2) In a generalized sense, for a steady non-ideal detonation the *sonic limit* can be designated as CJ surface since the true CJ condition can also be formulated as "$M = 1$". This surface, however, is not a plane; the term sonic limit is therefore preferable.

3) For a non-ideal detonation only the *frozen sonic limit* $M_f = 1$ can be of concern while the true CJ condition is to be described as *equilibrium sonic limit* $M_e = 1$.

4) In a non-ideal detonation the reaction is *not* completed at the sonic limit. In many cases, however, the completion of the reaction is achieved approximately at the sonic limit. As stated before, reactions which proceed beyond this limit have no influence on the detonation velocity. Moreover, due to the fast expansion behind the sonic limit these reactions will mostly freeze very soon without leading to chemical equilibrium, cf. [110].

5) It is correct that a certain *generalized CJ condition*, which has to be satisfied at the sonic limit, can be understood as determining the automatic adjustment of a certain stable non-ideal detonation velocity.

This last statement requires a more detailed motivation. For this purpose we consider a thin streamtube so that the process becomes *one-dimensional* and therefore can be described by functions which only depend on a single space variable. We take this variable x along the possibly curved axis of the stream tube in the downstream direction. The shock front may be at $x = 0$. The cross-sectional area of the stream tube $A(x)$ is for a non-ideal detonation a monotonically growing function of x. An actual physical meaning is associated only with the relative values $A(x_1)/A(x_2)$ or $A/A_0 = A(x)/A(0)$*. The *limit*

$$\varepsilon = \frac{1}{A}\frac{dA}{dx} = \frac{d \ln A}{dx} \tag{29.5}$$

formed from this ratio describes the local behavior (divergence or convergence) of the stream tube. In [110] ε has been introduced as a *expansion coefficient*. With ε according to (29.5) and Σ according to (5.28) the equations of motion for the flow in the stream tube can be written as

$$\frac{du}{dx} = -\frac{v}{u}\frac{dp}{dx} = \frac{u}{v}\frac{dv}{dx} - \varepsilon u = \frac{\varepsilon u - \Sigma}{M_f^2 - 1}, \tag{29.6}$$

$$\frac{dx}{dt} = u, \quad \frac{d\xi_j}{dt} = r_j. \tag{29.7}$$

This system of equations replaces the system which was established in section 8 for Rayleigh processes valid for ideal detonations. The system, however, is complete only if, besides specifications about the equation of state (4.2) and the reaction velocities $r_j(p, v, \xi_j)$, an additional determining equation for $\varepsilon = \varepsilon(x)$ or $A = A(x)$ is given.

Instead of using the differential equation (29.5) we could have written the finite equation

$$A\frac{u}{v} = \text{const} \tag{29.8}$$

which replaces the former continuity equation (1.1). We prefer the differential form since only ε, not, however, A, is a flow variable which is well defined for each point.

For a true *streamtube flow*, that means with practically constant values for all flow variables over each cross-section, the dynamic behavior of the casing-material will allow us to establish a differential equation for $A(x)$ where especially p occurs as an important quantity. The condition of stream-tube flow would probably be satisfied for not too large charge diameters and heavy

---

* In [108] the quantity A_0/A is denoted by θ, in [53] by C.

casing and best for gas detonations in (metal) tubes. In general, the conditions for stream-tube flow are not satisfied by a non-ideal detonation as a whole.

We may assume, highly simplifying, a stream-tube flow for the entire flow in cases of sufficient heavy casing or for the central part of the flow, as done in [109].

In an exact analysis, however, the interaction of the many different stream lines leads to a more-dimensional problem which can only be described in terms of partial differential equations. These, by the way, can be written in a form so that in addition to the relations (29.6), (29.7) valid for each stream tube a formula is obtained which describes the curvature of the streamtubes by means of the pressure gradient perpendicular to the flow direction. These formulas then determine the streamtube pattern and thereby also ε.

This determination of $\varepsilon(x)$ along each stream tube just as the one for $x(t)$ and $\xi_j(t)$ by (29.7) does not lead to singularities. On the other hand, the differential equations (29.6) become *singular* at the sonic limit $M_f = 1$. This behavior is familiar to us from the former discussed special case $\varepsilon = 0$, cf. (8.6).

For a streamtube in the steady non-ideal detonation—which we consider to be existent based on experimental proof—we can make certain general statements about the development of the flow variables, particularly p and u, inferring from (29.6).

Immediately behind the shock, at $x = 0$ and $t = 0$, p has large values, $M_f < 1$, and u is of some intermediate magnitude. Far downstream, for $x \to \infty$ and $t \to \infty$, p is small, u has large values, and $M_f > 1$. The flow passes a *sonic limit point* where $x = x_s$, $t = t_s$ and $M_f = 1$, and we assume that only one such point exists. The flow then proceeds mainly accelerated. It is therefore predominantly

$$\varepsilon u - \Sigma < 0, \quad M_f < 1 \quad (x < x_s,\ t < t_s) \tag{29.9}$$

or

$$\varepsilon u - \Sigma > 0, \quad M_f > 1 \quad (x > x_s,\ t > t_s). \tag{29.10}$$

For the transition through the sonic limit according to (29.6)

$$\varepsilon u = \Sigma, \quad M_f = 1 \quad (x = x_s,\ t = t_s) \tag{29.11}$$

must hold since du/dx remains finite.

We know from the discussion in Section 14 that for *non-pathological* explosives the reaction process is predominantly *expansive*, and this certainly at the end of the reaction, that means $\Sigma > 0$. At the beginning perhaps energy-consuming, *contractive* reactions with $\Sigma < 0$ may occur. At the end of the reaction, for $t \to \infty$, Σ approaches zero.

It is $\varepsilon > 0$ for $x \geq 0$. For the curved shock intersected perpendicularly by the considered parallel stream-tube flow immediately behind the shock [110]

$$\varepsilon = K\left(\frac{\varrho_1}{\varrho_0} - 1\right),$$

where ϱ_1/ϱ_0 is the condensation ratio of the shock and K the *mean curvature* of the shock front. It is $K = r^{-1}$ for cylindrical and $K = 2r^{-1}$ for spherical geometry with r as the radius of curvature.

It turns out that (29.9) may be violated behind the shock since $\varepsilon u > 0$ and $\Sigma \approx 0$ (induction period!) and that even $\Sigma < 0$ may be. The flow behind the shock then starts as a *condensation* which was also possible in case of an ideal detonation. At least near the sonic limit, however, (29.9) and (29.10) must hold since M_f is increasing there*.

Relation (29.11) is the initially mentioned *generalized CJ condition*. This condition has already been given in [108] p. 90, with reference to a report of DEVONSHIRE (1943). We have followed here, however, a more recent work [112] which distinguishes between a_e and a_f while [108] uses only one general velocity of sound.

29.4 *Stability of the detonation velocity*

By means of a t–M_f diagram, see Fig. 29.3 and compare also Fig. 14.6, we can see that a real non-ideal detonation which is characterized by Eqs. (29.9) to (29.11) is the only stable process compared with reactive shocks which propagate faster or slower.

The reaction zone in the expanding flow behind a normal shock which we consider as steady represents the solution of the system of ordinary differential equations (29.6) and (29.7). The initial values for this system correspond to the shock state and depend on a single *parameter*, the shock pressure or the shock velocity. Therefore ∞^1 integration paths result.

The argumentation now proceeds essentially as for the pathological detonation, cf. Section 14, and the topological characteristics of the integral curves in the t–M_f plane are here, cf. Fig. 29.3, the same as before, cf. Fig. 14.6. The former variable Σ is now replaced by $\Sigma - \varepsilon u$.

---

* That $M_f = u/a_f$ increases while u decreases, that p therefore is growing while a_f falls off can be assumed as very unlikely or *pathological*.

The integral curve of the steady non-ideal detonation, curve I in Fig. 29.3, passes the saddle point S as a *separatrix*, S being the intersection point of the curves $\Sigma = \varepsilon u$ and $M_f = 1$. For a stronger front shock smaller M_f values result (II); M_f remains always smaller than one. For weaker front shocks the curve runs higher (III) and terminates with *choking* at $M_f = 1$. Shocks propagating to fast are decelerated, slower shocks accelerated. The steady solution is *stable* against disturbances of the detonation velocity.

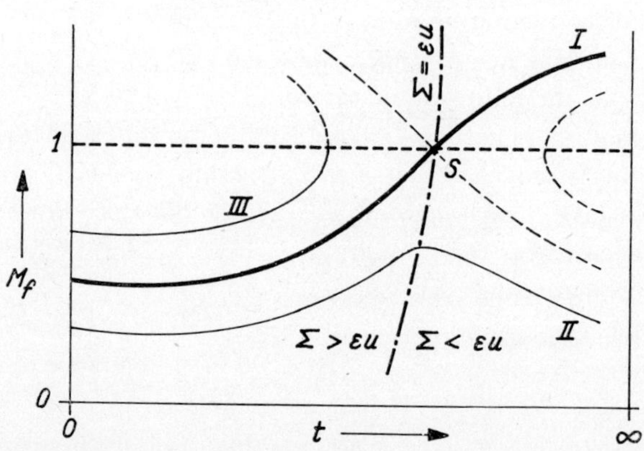

Fig. 29.3 Stability of the non-ideal detonation.

It is remarkable that the non-ideal detonation described in Fig. 29.3 corresponds to the *pathological* ideal detonation as shown in Fig. 14.6, not, however, to the normal ideal detonation (Fig. 14.2). The formation of a *quasi-Laval nozzle* here occurs without difficulty through the combined action of the geometrical divergence ε and the *thermal contraction* Σ/u, while the pure thermal nozzle with a throat for $\varepsilon = 0$ represents a pathological, exceptional case.

Formally, the steady streamtube flow with reaction and variable cross-sectional area presents scarcely new possibilities as compared with the special case of constant cross-sectional area, which is pointed out in [129]. But the actual behavior of the substances known to us implies essential differences.

We therefore can see how for given casing conditions, that means for a somehow prescribed $\varepsilon(x)$, a certain shock intensity and therefore a detonation velocity D together with a particular integral curve and a reaction-zone structure is established. In a real two- or three-dimensional steady non-ideal detonation each single stream tube must satisfy independently the condition (29.11).

If $\varepsilon(x)$ is given explicitly D can be in principle calculated by means of (29.11). But any prescription of $\varepsilon(x)$ is certainly always arbitrary since in reality this function is determined by the interaction of the different stream tubes. $\varepsilon \equiv 0$ corresponds to the ideal detonation. Larger ε means stronger deviation from the ideal case and therefore lower strength of the casing material.

In order to obtain a relation between casing conditions and detonation velocity, $\varepsilon =$ const has been assumed in [110] which is a strong simplification. In this way a certain function $D = D(\varepsilon)$ was found for a certain explosive.

Since the *stable* non-ideal detonation according to Fig. 29.3 represents the limiting case between choking and pure subsonic solutions the points with stable detonation lie in the D–ε plane on the limiting curve between a *choking* and a *subsonic* region. In [110] it is proved that normally the limiting curve does not run monotonically, but rather a descending and an ascending curve branch meet at a point where $\varepsilon =$ maximum. It is further shown there that only the descending branch corresponds to the stable detonation while the ascending branch is a *sensitivity curve*, and that the detonation fails when the point $\varepsilon =$ max is reached.

29.5 *Applications*

The quantity ε is mainly important for hypothetical experiments. For practical cases the charge radius R or the casing is determining for the *degree of non-ideal behavior*. For shock waves of various intensities (D) in charges of various diameters ($2R$) the same alternative exists as for the idealized one-dimensional process, namely between selfacceleration (choking) and deceleration (subsonic flow).

Correspondingly limiting curves in a diagram with coordinates D/D_i and l_i/R (subscript i for ideal behavior) are shown in Ref. [108], p. 134. These curves show, as long as the dependence $l = l(R)$ is regarded, the previously mentioned non-monotonic behavior.

The theory has been improved in [111] and was applied to two well known substances, cf. Fig. 29.4. These substances are liquid TNT and ammonia nitrate with crystal density. The theory requires values for the radius of curvature s of the front shock and the calculation has been carried out with different values of s/R. Each of the curves has a point $R =$ min. For this limiting radius the detonation velocity is only a few percent below its ideal value. Its variability is therefore small. Above the curve, that means for better casing, acceleration of the shock is obtained, below the curve deceleration. Consequently the righthand branch of the curve, ascending in this diagram, is *stable*. The other, *instable* branch is to be considered as a *sensitivity curve*. Points which are on the lefthand side of this branch are associated with shocks which are too weak in order to initiate a sufficient reaction for the particular radius *R*. Those shocks decay like ordinary shocks.

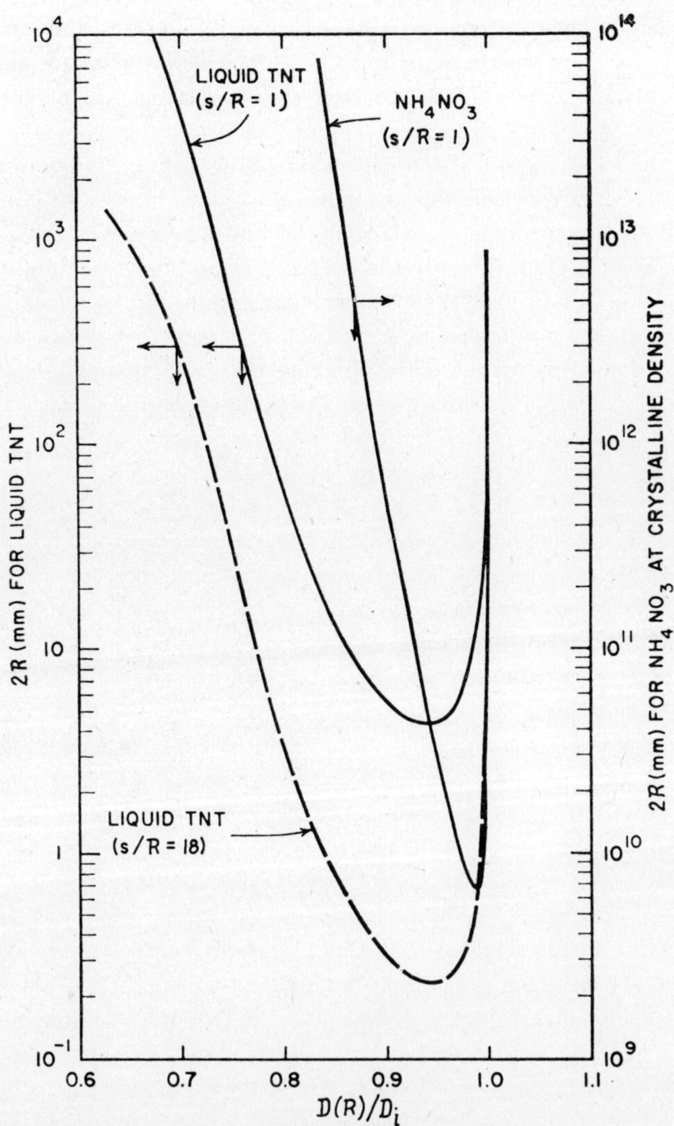

Fig. 29.4 Limiting curves for two explosives (from Ref. [111)].

The resulting limiting radius for TNT at $s = R$ is 2.1 mm. An experimental result is 6 mm. Regarding the simplifications and uncertainties of the theory the agreement of the values is satisfactory. For NH_4NO_3 the enormous value of 3300 km is found, which means that this substance is not detonable under any realistic circumstances; and this is indeed in accordance with experience*.

This theoretical result proves our remark made at the end of Section 15. For substances which satisfy condition 1, listed there, a statement about the detonability gets a clear meaning only if at least an order of magnitude of a limiting radius or perhaps a reaction zone length is given in addition.

While the data listed in Table 16.1 regarding the characteristic values of the CJ detonation are exclusively determined by thermodynamic facts and are therefore unsensitive against small *admixtures* as well as changes of the microcrystalline structure, the reaction zone structure, its length, and the limiting radius show a quite different behavior. The crystalline structure as well as certain additives in small amounts may well change considerably the detonability and that means just the reaction zone length or the reaction rate, cf. [113]. Additives increasing or decreasing the *sensibility* are of course of great practical importance.

A quite surprising behavior for solid explosives was found when systematic variations of the charge diameter, the explosive type, and the charge density were made [137]. While the ideal detonation velocity, as is known for a long time, always increases approximately linearly with the charge density, the substances could be divided into *two groups* regarding the *diameter effect.* Substances of *group 1* show with increasing charge density, that is with decreasing porosity, more and more ideal behavior, which means decreasing reaction zone length. These are in general the *high-energetic explosives.* Substances belonging to *group 2* behave contrarily. It seems that for the second group the generation of *hot spots* supported by *cavities* is decisive for the ignition, while for the first group owing to the higher pressures always enough ignition points are produced in sufficient dense distribution. The higher energy density connected with the greater charge density now becomes profitable for the ignition. Correspondingly, a *homogeneous* reaction type is associated with the first group and with the second group a *heterogeneous* reaction type. It should be noticed, however, that for both groups the reaction occurs as a grain deflagration and therefore as a two-phase process (which is heterogeneous in the common sense, cf. Section 11.4).

---

* NH_4NO_3 with diminished density is quite detonable; it belongs to the second group of explosives according to Ref. [137]; see Fig. 29.5b.

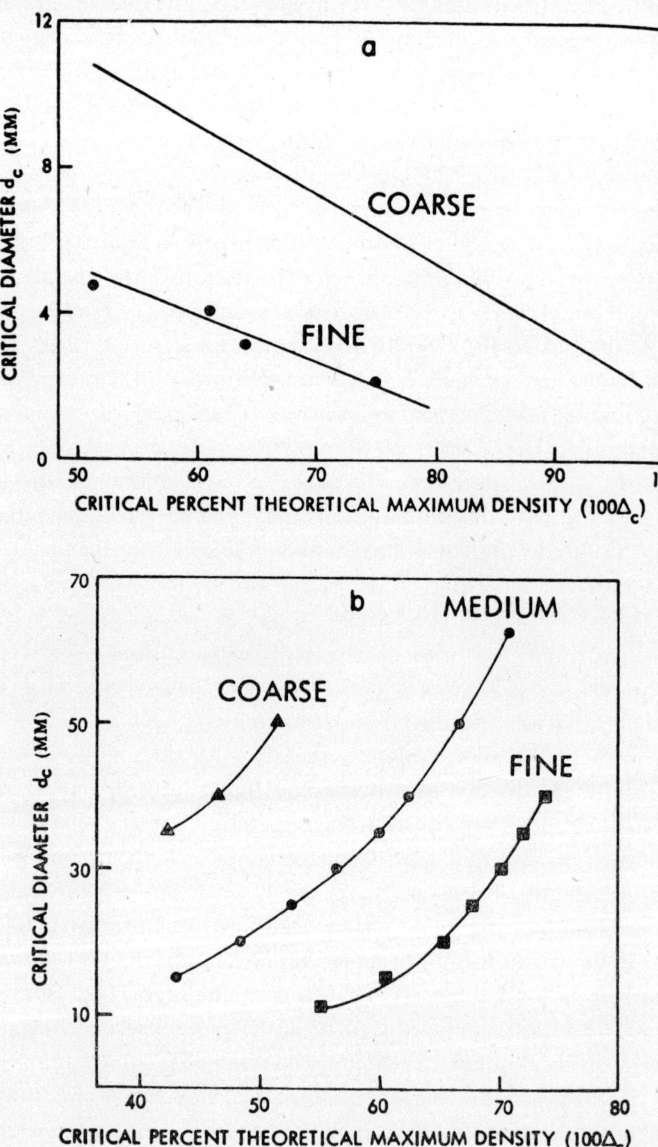

Fig. 29.5 Different behavior of the explosives from group 1(a), and group 2(b) (from Ref. [137]).

In Fig. 29.5 the detonation limits are shown for typical representative samples from both groups in the Δ-d diagram. Δ is here the charge density divided by its maximum value for perfect space occupation; d is the charge diameter. The critical charge diameter changes in the opposite sense as the charge density for group 1 (here TNT) but in the same sense for group 2 (here ammonia nitrate with 12% peat meal). Simultaneously the grain size is here also varied. For the grain-deflagration model the reaction-zone length and thereby the critical diameter are both approximately proportional to the grain size for a given explosive. In this respect both groups behave likewise.

Certain liquid and gelatine explosives are able to detonate with two distinctively different velocities under certain conditions [114]. The greater one of the two velocities is in the ideal case the CJ value. The smaller one is only slightly higher than the velocity of sound. This *low-velocity detonation* appears still to be somewhat obscure according to the more recent work [135]. Obviously only a small fraction of the chemical energy released in the ideal detonation becomes available. We may ask:

1) What is the mechanism which enables the reaction to be initiated by the relatively weak shock?

2) What prevents the approach of the chemical equilibrium, that means the complete release of the entire energy in the reaction zone?

The existing work has predominantly investigated the first question and has clarified it fairly extensively.

The second question will probably be answered mainly by pointing out that the slow detonation exists only as a non-ideal detonation. Apparently the energy-producing reactions here occur mainly beyond the sonic limit. In [110] it was suggested as a condition for the existence of the slow detonation that two extrema of the discussed limiting curve appear, cf. Fig. 29.4, resulting in two separate stable branches. A curve of such a shape, however, has not yet been established in any case.

Reference [135] points out that slow detonations of thin liquid films may occur even for rigid casing and that therefore the lateral expansion cannot be decisive. But as we shall see immediately, also for a rigid casing a true diameter effect and a non-ideal behavior is possible.

For detonating gases, in contrast to condensed explosives, a practically rigid, inflexible casing can be made fairly easy. We therefore should expect a quasi-steady detonation of a gas which is initially at rest in a long cylindrical tube of sufficient strength (to withstand the detonation pressure), to proceed as an ideal detonation with D_{CJ}. This, however, is not correct. Rather also in this case a similar diameter effect is observed as for solid explosives with lateral expansion. This property is explained as boundary-layer effect in [115]. Since the solid wall material keeps moving exactly with the approaching-stream velocity D, while the gas, however, is decelerated considerably by the shock transition, a boundary layer with higher velocity developes along the tube wall detracting matter from the mainstream. The negative *displacement thickness* $-\delta^* < 0$ of the boundary layer is equivalent to an effective enlargement of the tube from R to $R + \delta^*$. In the main center flow actually a lateral expansion occurs and the theory of [108] etc. is applicable. The quantitative analysis of this idea leads to a satisfactory agreement with the experimental results. Reference [115], by the way, uses a *chemical relaxation length* instead of using a reaction-zone length. Thereby, the hypothesis of a strictly bounded reaction zone is avoided.

If projectiles are fired with high velocities into an explosive gas mixture, a steady* flow pattern with shock front, reaction zone and lateral expansion is obtained [116]. A typical configuration is shown in Fig. 29.6. The velocity of the projectile is greater than the ideal detonation velocity D_{CJ} of the gas mixture, see (25.20′).

To each point on the shock front belongs a certain shock Mach number M_s or normal component D of the approaching-stream velocity as well as an expansion coefficient ε. ε can be determined from the curvature, the inclination, and the condensation ratio of the shock. If we assume ε to be constant along the streamlines close behind the shock we obtain approximately the conditions as assumed in [110].

The shock which is intense near the axis immediately initiates the reaction and may be described as a strong detonation. Proceeding towards the exterior parts of the flow, however, a transition into a self-supporting detonation does *not* occur. The shock further decays after the flame front has separated and eventually becomes a *Mach line*. Nowhere a *choking* occurs. The repercussion of the reaction on the shock front is small. The image curve of the shock line in the D–ε diagram remains entirely outside of the limiting curve. It is therefore appropriate to use the term *shock-induced* combustion rather than detonation.

A considerably larger sphere under equal circumstances would have

---

* Instabilities also investigated in [116] will not be discussed here; compare regarding this aspect the remarks made in Section 13d.

initiated without doubt a self-supported detonation. The determining gas-dynamic and reaction-kinetic conditions for this effect are discussed in [117], and the required sphere diameter is estimated.

In a comprehensive research work especially conducted at the University of Michigan [118]–[121] the non-ideal detonation in gases is investigated. A column of detonable gas is separated there from a surrounding inert gas by

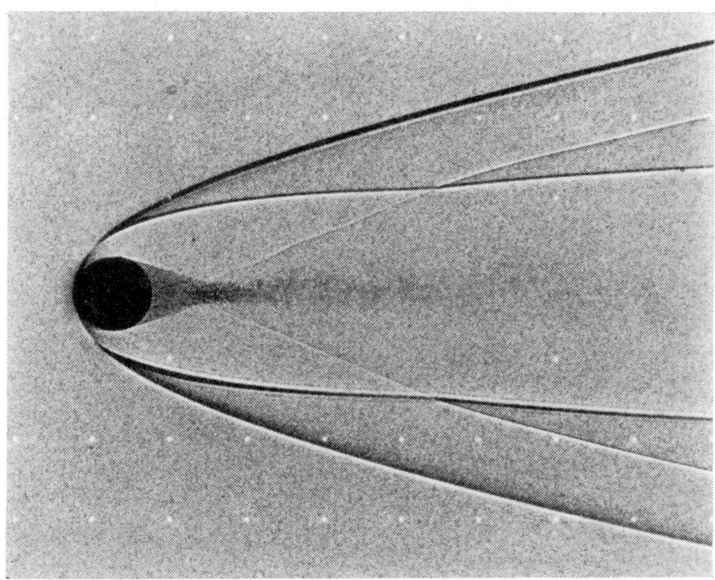

Fig. 29.6 Bow shock wave and combustion front generated by a polycarbonate sphere (diameter 9 mm) fired into a stoichiometric hydrogen-air mixture ($p = 0.55$ atm, $a = 405$ m/s, $D_{CJ} = 1950$ m/s). Velocity of the sphere 2870 m/s. Ablation products are visible in the wake. (From Ref. [116].)

a very thin membrane. The inert gas acts as a highly compliant casing. Experiments with detonating gases are considerably simpler than those with detonating condensed explosives. The physical processes, however, are similar in both cases, and the experiments are therefore very instructive for the non-ideal explosive detonation. Moreover, the diameter effect which depends on the reaction-zone length permits to draw conclusions about the reaction kinetics. In this way a catalytic effect of water vapor on a detonating oxyhydrogen gas is proved in [121].

30 Unsteady Non-Ideal Detonations

30.1 General characteristics

The propagating detonation which appears primarily as an unsteady process for an impartial observer has at the same time the basic physical property to propagate with constant velocity in an unlimited explosive. Therefore, it is best described as a steady flow phenomenon. If, due to the geometry of the propagation or the flow pattern, unsteady processes were considered within the scope of ideal detonations, the *Hugoniot jump conditions* could still be applied to each part of the detonation front. These conditions, however, are in principle valid only for the plane steady detonation wave, as well for the CJ detonation as for a strong detonation.

After we have considered only steady processes also for the non-ideal detonation, an important complex of problems is left which governs the origin and the extinction of detonation, in other words, the transient phenomena. These are in principle unsteady, non-ideal phenomena and are to be treated as those.

It is evident that in order to control detonation processes a good understanding of the transient processes is very desirable. For a long time these phenomena were studied only from a technical-practical or at most experimental-physical point of view. The progress made in physico-chemistry and in the computer sciences has now made possible also a certain theoretical-mathematical treatment of these problems.

Before passing on to the discussion of related attempts, we shall make at first an addition to Chap. V. We have shown there that a reactive shock is accelerated or decelerated if its velocity D is smaller or larger than D_{CJ} respectively. But we were not able to give explicitly the positive or negative *shock acceleration.*

Assuming certain approximations in [108] a relation

$$\frac{dD}{dx} = \frac{K}{l}(D_{CJ} - D) \tag{30.1}$$

is obtained which closes this gap. x is the space coordinate in the propagation direction, l the reaction-zone length, and K a dimensionless constant dependent on the equation of state and the reaction energy ([108] gives $K = 0.333$).

It is remarkable that this formula, which has been proved also experimentally, comprises both cases $D \lessgtr D_{CJ}$ which in the quasi-steady discussion, cf. Ch. V, appear as qualitatively different. Indeed, the idea of thermal choking used there has no sense in an unsteady treatment since the condition $M = 1$ is not *coordinate-invariant.*

In addition to (30.1) an analogous relation for the asymptotic adjustment of the stable non-ideal detonation velocity D_{st} in a charge with finite diameter is obtained in [108], namely

$$\frac{dD}{dx} = \frac{K}{l}(D_{st} - D). \tag{30.2}$$

The constant K is the same as in (30.1). Both formulas are applicable for reactive shocks which differ only slightly from the stable detonation. The formulas do not relate to the formation of the detonation process.

30.2 *The formation of detonations*

In technical applications the bulk explosive charge is always initiated by means of a detonator containing a *primary explosive.* This is in general *initiated mechanically* (by impact) or *electrically.* In the latter case energy is added mainly in form of heat.

Primary explosives are qualitatively different from the *secondary explosives* with respect to their chemical structure as well as to their detonative behavior. The former immediately detonate when heat is added and a certain ignition temperature is reached without exhibiting any preceding nearly-steady deflagration. They can be characterized as materials capable of a detonation but not of a deflagration.* This behavior may depend, of course, not only on the chemical structure of the substance but also on its physical state, particularly its density.

We therefore have essentially two mechanisms of *detonation initiation*:

a) Initiation by means of *heat addition* (possible only in primary explosives), and
b) Initiation through *shock waves* (possible for all explosives).

Heating and shock may of course be active simultaneously. Other mechanisms like radiation or chemical effects are fairly unimportant besides a) and b).

Between the small primary charge and the secondary or main blasting charge which may be very large and unsensitive very often one or several

* The demonstrated steady slow combustion of very small and thin monocrystals of typical primary explosives [132] is obviously possible only with a continuous lateral heat loss but not as a one-dimensional process or in a compact substance.

booster (transmission or amplification) charges are placed in form of a chain with graduating amounts. These are secondary explosives of gradually decreasing shock sensitivity yielding a better matching for the transmission of the detonation process.

The frequently used direct transmission of a detonation to an adjacent second explosive belongs to the mechanism b) mentioned above since the causing effect is only the shock wave*.

There is no essential difference whether the "*acceptor charge*" and the "*donor charge*" are in direct contact or an inert material is placed inbetween which only transmits the shock.

We must carefully distinguish the initiation mechanism a) from the spontaneous *transition* of a combustion or deflagration into a detonation. This latter phenomenon was studied particularly in gases but also in solid explosives [122], [123]. Since a nearly stationary propagating flame front is assumed, primary explosives are excluded. The flame which produces a volume increase in propagating with subsonic velocity generates pressure waves in the forward and backward direction. These pressure waves may steepen to shock waves and may eventually lead to detonation, sometimes by interaction with inhomogeneities and boundaries of the medium, sometimes at the original flame front after reflection from walls. For gases, *turbulence* in the flow generated by pressure waves may also be of influence. This is obviously a complicated flow process with reactions, including essentially shock wave effects while the combustion itself still might well propagate as a steady process.

30.3 *Theoretical-mathematical investigations*

The direct initiation through heat increase has apparently not been investigated theoretically. The one-dimensional, unsteady propagation of *reactive shocks*, on the other hand, has been analyzed mathematically-numerically in several cases.

In [124] a square-shaped pressure pulse of certain intensity p_1 and duration t_1 is assumed to be generated at the surface of a semi-infinite explosive space. Based on simple assumptions for the equation of state and the reaction kinetics (Arrhenius formula) the space-time development of the state results from a numerical integration using the *von Neumann–Richtmyer Method.* Depending on the values of p_1 and t_1 detonation or failure results. The limit

* The idea that transport phenomena are essential in a steady detonation front as well as for the detonation transmission [114] has not been accepted generally.

between these cases is fairly sharp. In two cases with $p_1 = 10^5$ atm and $t_1 = 0.69$ and $0.64\ \mu s$, respectively, it turns out clearly already for $t = 1\ \mu s$ that the first calculation yields a detonation, the second, however, not.

Accordingly a *limiting curve* could be drawn in a p_1-t_1 diagram which separates the initiating shocks from the non-initiating shocks. In [124] instead, however, the question is raised how long at a certain point in the explosive a certain energy density E_0 resulting from shock compression must be maintained in order to initiate the detonation at that point. The result is a limiting curve in a E_0–t diagram, cf. Fig. 30.1. Four curves are shown in the

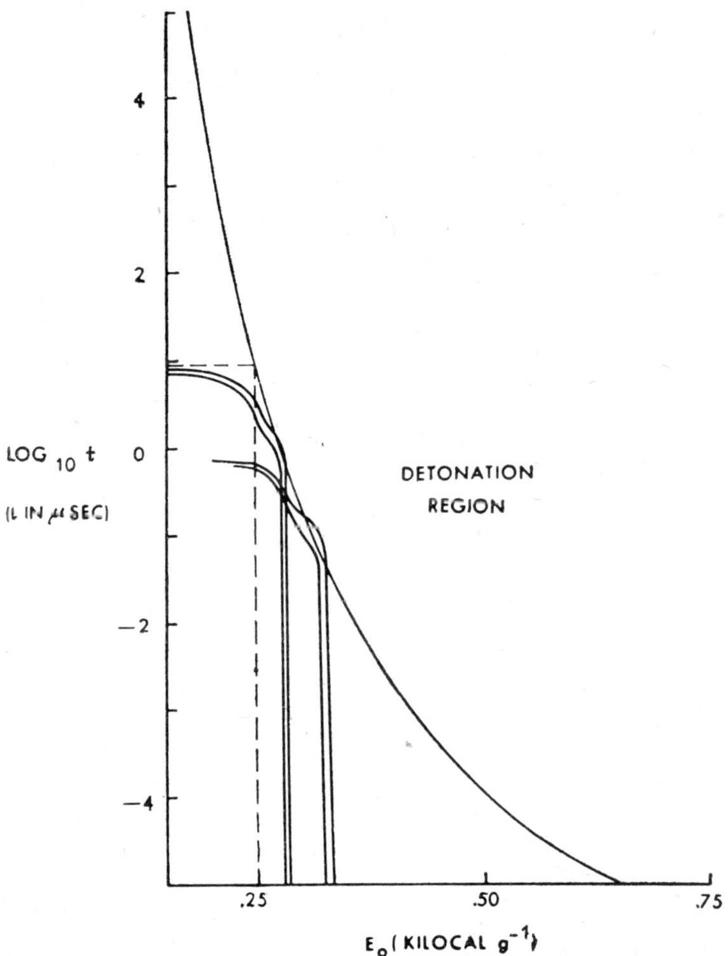

Fig. 30.1 Limiting curve for detonation initiation (from Ref. [124]).

same diagram which describe the effective duration of different E_0 values at certain points for two cases. They partly transgress the limiting curve.

This curve in the E_0–t plane has a close relationship to the limiting curves mentioned in 29.4 and 29.5 where besides a shock-intensity parameter the expansion coefficient ε or the charge radius R has been used as a second coordinate. The longer effective duration corresponds to the lower expansion or the larger charge radius. Only, however, the sensitivity branch (the descending branch in Fig. 29.4) has a certain correspondence in the E_0–t plane.

In a similar investigation [122], [123] the starting configuration is a rigidly cased plane explosive surface from which a combustion starts. There the pressure increases rapidly interacting with the reaction. Pressure waves are generated which propagate into the medium accumulating into shock waves which then initiate the detonation. [122] describes the experiments which are analysed numerically in [123]. The at first slowly propagating flame front is here, in fact, simulated by a certain pressure variation at $x = 0$. Again, the detonation starts at a certain distance from the surface, e.g. 15 cm in one case. The fact that the pressure values required for the detonation initiation were found to be smaller in the experiment than in the calculation is explained by randomly distributed hot spots in the real explosive which are not taken into account in the mathematical model. Otherwise, experiment and calculation are in good agreement.

References [125], [126] also describe parallel conducted experiments and calculations. An intense shock wave in the explosive is generated here by the impact of a high-velocity metal plate. The calculation uses as realistic as possible assumptions for the boundary conditions, the equations of state, and reaction-kinetics. Particularly a two-phase grain deflagration is assumed with in general different temperature for the explosive and the combustion products.

In Fig. 30.2 the pressure development resulting from the impact of a copper plate is shown. It can be seen that after a time $t \approx 1.7$ μsec the detonation starts at a distance $x \approx 7$ mm propagating in both directions. The backward propagating wave is strongly attenuated since it runs into an already considerably reacted medium.

Reference [134] attempts to describe analytically the experimentally observed differences in the behavior of liquid and solid explosives by means of a proper model. In both cases two phases are assumed, an unreacted condensed and a reacted gaseous phase. For the reaction velocity two different arrangements are used. The results, however, make the applicability of the model questionable.

While these investigations, in spite of the heterogeneous reaction, have presumed a one-dimensional flow, Refs. [127] and [128] treat the interaction of a shock wave with a single, spatially limited inhomogeneity of the explosive. Hereby the origin and effect of *hot spots* become understandable. The interaction of a shock with a disturbance point in nitromethane is

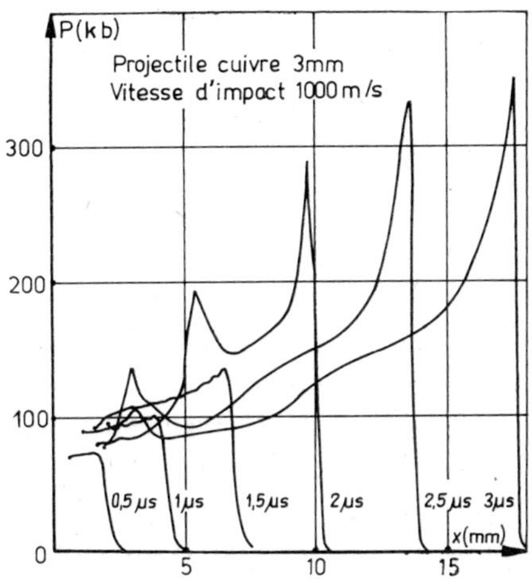

Fig. 30.2 Detonation initiation after impact (from Ref. [126]).

calculated as a two-dimensional unsteady flow problem by means of the PIC-(Particle-In-Cell) method. Depending on the different conditions detonation or failure results. Without disturbance spots either detonation failure results or at least a longer transient period (distance) is necessary until the detonation arises.

The time-dependent reaction zone behind a plane shock is calculated in [31]. The shock is supposed to be generated by a piston moving towards the explosive (Nitromethane). The steady structure of an overdriven detonation is gradually established. But even when the conversion is completed already at the piston surface the initiated detonation can still go out again through a sudden strong reduction of the piston velocity.

The integration method of [31] is also applicable to cylindrical and spherical propagation. These cases are not treated in [31], but* the spherical shock propagation has also been calculated, yielding the approximate initiation energy necessary for the generation of a stable detonation.

For the ideal self-similar *spherical* detonation we have found in Section 23 a

* According to private communication.

steep pressure decrease immediately behind the detonation front. This behavior must be approached asymptotically in a real spherical detonation even if an intense primary charge provides a good starting of the detonation. It seems that the mentioned pressure decrease disturbes the reaction zone and may even lead to an extinction of the detonation having already started. So a large charge of TNT which has a long reaction zone when initiated at a point detonates completely only if a sufficient large *booster charge* with short reaction zone like RDX or PETN is used.

30.4 *Physical investigations*

We cannot enter here a detailed discussion of the many experimental investigations of the unsteady non-ideal detonation. We rather restrict our outline giving some references and listing a few important aspects.

We consider particularly initiation processes which lead to a detonation either directly or through an intermediate step comprising a quasi-steady deflagration. In order to differentiate we may call the start of a pure deflagration without appreciable pressure increase *inflammation* or *ignition*.

The initiation problem has been treated in monographs [131], [132] as well as in conference proceedings e.g. [133], [130], [103], [105].

We should distinguish between physical and technological treatments though a limit cannot be drawn exactly between these two approaches. The investigation of the generation and transmission of detonations was governed for a long time by technological procedures dictated by the practical requirements. A physical-theoretical analysis had to be abandoned mostly. In the last decades progress has been made in defining and formulating the physico-chemical processes governing the complex phenomenon. Additional improvements, however, are still desirable.

The contributions on initiation processes in [103] give a fairly representative survey of the modern research in this field. In a review article (pp. 512 to 516) these are divided into four groups discriminating between *physical-experimental*, *mathematical-theoretical*, and *technological* work and, concerning the first, whether the aim is more a qualitative and fundamental description of the phenomena, or quantitative and applicable determination of the physical quantities governing inflammation and initiation.

Regarding the bipartition of the physical causes leading to an initiation in Section 30.2, namely heating and shock influence, we remark that for the beginning of chemical reactions always the temperature increase is decisive which can be a secondary effect. For the initiation of a detonation, however, the effect of an intense, perhaps secondary, shock is determining.

Practically the heat may be supplied by conduction or radiation, and may be produced electrically through heat wires, spark discharge, wire explosion, or mechanically through impact, shock waves, or friction.

The *continuing ignition* in a steady reactive shock wave, that means shock-induced combustion, ideal or non-ideal detonation, represents a special case of the pure shock initiation. This again is a special case of initiation processes with different agents. The ignition problem touched briefly in Section 11 can be investigated systematically only as an unsteady problem since the determining parameters as pressure, temperature, and duration can be varied appropriately only in this case.

Regarding the gasdynamic theory of detonation we are predominantly interested in the shock initiation. The homogeneous shock-adiabatic heating is, however, decisive for the reaction start in gases only. For condensed explosives the energy increase through the shock is only moderate due to the low compressibility, and the gain in the overall temperature is not sufficient to induce the reaction, at least without too long induction period. The generation of hot spots where the reaction starts is then decisive; see Sections 11, 29.5, 30.3. Very small gas bubbles in a liquid explosive increase considerably its shock sensitivity since through pressure jumps of certain intensity much higher temperatures are generated in the gas than in the liquid. Each small bubble acts as an ignition point.

In solid explosives hot spots can originate in the same way as described above or through friction and fracture within the crystalline structure. Strongly increased pressures and temperatures hereby locally occur, and that more likely for a certain explosive at lower densities, most unlikely in monocrystals.

The difficulty of the initiation problem is in general mainly due to the physical and chemical inhomogeneity of the participating substances. The behavior on shock loading and heating is therefore very difficult to describe.

In addition to the variety of explosives and their mixtures with other explosives or inert materials there exists a multiplicity of various agents. Also the requirements for the ignition behavior are numerous, e.g. sensitivity with respect to certain agents, insensitivity against others, brevity or precision of the sensation period.

We therefore conceive that just the unsteady non-ideal detonation processes represent an extended research topic.

Symbols

Formula numbers or page numbers indicate where the symbol is introduced first in the text.

a	sound velocity (2.26)
	(also used as a coefficient in formula (19.1))
a_e	equilibrium sound velocity (4.7)
a_f	frozen sound velocity (4.8)
A_j	affinity (20.8)
$A(x)$	streamtube cross sectional area, p. 165
b	co-volume (eigen-volume) of molecules (19.1)
B, C	virial coefficients (19.2)
$\mathscr{B}$	material boundary, p. 128
c_p	specific heat capacity at constant pressure (2.20), (5.16)
c_v	specific heat capacity at constant volume (2.19), (5.16)
C_p, C_v	molar heat capacities, p. 12
D	detonation front velocity, p. 51
$\mathscr{D}$	detonation polar or detonation wave, p. 122
E	specific internal energy, p. 5
$\mathscr{E}$	expansion (rarefaction) wave or polar, p. 129
E_{ch}	chemical energy (9.2)
f	number of quasi-degrees of freedom (2.21′)
F	specific free energy (2.9); also, heat parameter (9.19)
G	specific Gibbs enthalpy (2.10)
H	specific enthalpy, p. 2, 8
$\mathscr{H}$	Hugoniot curve, p. 21
$\mathscr{I}$	isentrope, p. 23
k	reaction-rate constant (11.1)
K	constant, (19.6), with different meaning also in (30.1)
K	(degree) Kelvin
l	reaction zone length, p. 162
m	mass (2.15)
$M = u/a$	Mach number (7.5)
M_s	shock Mach number (7.6)
M_{mol}	molar mass (2.15), p. 112
$\mathscr{M}$	Mach stem (shock), p. 156
n	mole number (2.15)
p	pressure (1.2)

p_i^*	fugacity (20.30)
p_r	formal rest pressure (1.2)
$\mathscr{P}$	polar curve, p. 155
q	differential reaction energy (5.9)–(5.11)
Q	reaction energy or heat added per unit mass (5.1)–(5.8)
$\bar{Q} = Q/c_pT_0$	Damköhler parameter (9.4)
r, r_j	reaction rate (4.3); also, distance from center of symmetry (23.2)
R	specific gas constant R_a/M_{mol} (2.16); also, radius (23.1)
R_a	universal gas constant (2.15)
$\mathscr{R}$	Rayleigh line (3.8)
s	Radius of curvature of the shock front, p. 169
S	specific entropy (2.5)
$\mathscr{S}$	shock, shock polar, p. 129
t	time, p. 6
T	absolute temperature (1.3)
u	flow velocity (x-component) (1.1)
U_M, U_p	material velocities, p. 133
U_S	shock velocity, p. 133
v	specific volume, p. 5
V	volume (2.15)
W	mechanical work, p. 92
W_c, W_e	mass of casing, of explosive (29.3)
x, x', y	Cartesian coordinates, p. 1, 6
y	Riemann function (23.14)
$z = r/t$	pseudostationary coordinate (23.2)
α	thermal expansion coefficient (2.29); also, used as angle of incidence, p. 134
α_M	Mach angle, p. 136
$\gamma = c_p/c_v$	specific-heat ratio or adiabatic exponent (2.21)
$\Gamma = a^2/pv$	generalized adiabatic exponent (2.27)
δ	deflection angle, Fig. 25.1
Δ	charge density, Fig. 29.5; difference symbol, p. 3
$\varepsilon(x)$	coefficient describing the change of streamtube cross-sectional area (29.5)
$\eta, \bar{\eta}, \hat{\eta}$	viscosity coefficients (1.2), p. 4
$\theta = \varrho u$	mass flux density (1.1); also, flow angle, Fig. 26.1
λ	coefficient of thermal conductivity (1.3)
μ	chemical potential, p. 30, (20.3); also, condensation ratio (25.8)
ν	dimension number (23.4)
ν^{j_i}	stoichiometric coefficient (20.4)
ξ	reaction variable, extent of reaction (4.1)
ϱ	density (1.1)
σ	coefficient of expansivity (5.25) also, shock angle, Fig. 25.1
σ_{xx}	stress component, p. 3
$\Sigma = \sum_i \sigma_j r_j$	expansivity (5.28)
τ	relaxation time (6.1)
ω	angular frequency, p. 35

Subscripts

J, CJ	Chapman-Jouguet (7.2)
e, eq	equilibrium, p. 34
f, fr	frozen, p. 34
i, j	number of reaction, p. 30
r	rest or stagnation state (1.2), (1.3)
p, v, T etc.	quantities kept constant (2.12)
0, 1, 2	flow states, Figs. 1.1, 4.1

Bibliography

A. Cited References

Chapter I

1 Courant, R., Friedrichs, K. O., *Supersonic Flow and Shock Waves*, Interscience Publishers Inc., New York–London (1948).

2 Sauer, R., *Nichtstationäre Probleme der Gasdynamik*, Springer-Verlag, Berlin–Heidelberg–New York (1966).

3 Fundamentals of Gasdynamics. Editor: H. W. Emmons. Vol. III of *High Speed Aerodynamics and Jet Propulsion*, Princeton University Press (1958).

4 Oswatitsch, K., 'Physikalische Grundlagen der Strömungslehre', *Encyclopedia of Physics*, Editor: S. Flügge, Vol. VIII/1, Fluid Dynamics. Springer-Verlag (1959).

5 Becker, E., *Gasdynamik*. B. G. Teubners Verlagsgesellschaft, Stuttgart (1966), English Translation, Academic Press (1969).

6 Hayes, W. D., Probstein, R. F., *Hypersonic Flow Theory. Vol. 1: Inviscid Flow*. Academic Press, New York–London (1966).

7 Zeldovich, Ya. B., Raizer, Yu. P., *Physics of Shock Waves and High Temperature Hydrodynamic Phenomena. Vol. I and II*. Edited by W. D. Hayes and R. F. Probstein. Academic Press, New York–London (1966).

8 Landau, L. D., Lifshitz, E. M., *Fluid Mechanics*, Pergamon Press (1959).

9 Vincenti, W. G., Kruger, Ch. H. (Jr.), *Introduction to Physical Gas Dynamics*, John Wiley & Sons, New York–London–Sydney (1965).

10 Clarke, J. F., McChesney, M., *The Dynamics of Real Gases*, Butterworths, London (1964).

11 Oertel, H., *Stoßrohre. Theorie, Praxis, Anwendungen (Shock Tubes)*, Springer Verlag, Wien (1966).

12 Becker, R., *Theorie der Wärme*, Springer Verlag, Berlin–Göttingen–Heidelberg (1955).

13 'Principles of Thermodynamics and Statistics', *Encyclopedia of Physics*. Edited by S. Flügge; Vol. III/2. Springer Verlag (1959).

14 Kirkwood, J. G., Oppenheim, I., *Chemical Thermodynamics*, McGraw-Hill (1961).

15 Wecken, F., *Mechanische und thermische Effekte chemischer Reaktionen in kompressibler Strömung*, Technische Mitteilung ISL–T 2/63 (1963), also Roy. Aircr. Establ., Farnborough, Libr. Transl. No. 1261 (1967).

16 Becker, R., 'Stoßwelle und Detonation', *Z. Physik* **8**, 321 (1922).

17 Döring, W., Burkhardt, G., 'Beiträge zur Theorie der Detonation', *Deutsche Luftfahrtforschung*, Bericht Nr. 1939 (1944).

18 SAUTER, F., 'Entropie- und Geschwindigkeitsverhältnis bei stationär laufenden Umwandlungsfronten', *DEFA/LRBA – Note Technique* **8a/47** (1947) (ISL).

19 Combustion Processes. Editors: B. Lewis, R. N. Pease, H. S. Taylor. Vol. II of *High Speed Aerodynamics and Jet Propulsion*, Section B. Princeton University Press (1956).

Chapter II

20 GREENE, E. F., TOENNIES, J. P., *Chemical Reactions in Shock Waves*, Academic Press (1964).

21 STUPOCHENKO, YE. V., LOSEV, S. A., OSIPOV, A. I., *Relaxation in Shock Waves*. Moscow (Russian, 1965), Berlin–Heidelberg (English, 1967).

22 BETHE, H. A., TELLER, E., *Deviations from Thermal Equilibrium in Shock Waves*, Cornell University (1941).

23 *Sélection de constantes physico–chimiques relatives aux propergols et à leurs produits de réaction;* C.E.P.A./9843 (1956).

24 LIGHTHILL, M. J., 'Viscosity Effects in Sound Waves of Finite Amplitude', *Surveys in Mechanics* (1956) 250–351.

25 KIRKWOOD, J. G., WOOD, W. W., 'Structure of a Steady-State Plane Detonation Wave with Finite Reaction Rate', *J. Chem. Phys.* **22**, 1915–1919 (1954).

26 BECKER, E., 'Eindimensionale, stationäre Verdichtungsströmungen in einem Gas mit Relaxation', *ZAMM* **46**, 363–376 (1966).

27 DÖRING, W., 'Die Geschwindigkeit und Struktur von intensiven Stoßwellen in Gasen', *Ann. d. Physik VI* **5**, 133–150 (1949).

Chapter III

28 ZELDOVICH, J. B., KOMPANEETS, A. S., *Theory of Detonation*, Academic Press (1960), originally in Russian (1955).

29 MULLINS, B. P., PENNER, S. S., *Explosions, Detonations, Flammability and Ignition*, Agardograph No. 31, Pergamon Press (1959), Chapter V.

30 FOREMAN, K. M., 'Pressure Variation of Strong Detonation Limits', *J. Aer. Sci.* **29**, 240, 469 (1962).

31 MADER, C. L., 'A Study of the One-Dimensional Time-Dependent Reaction Zone of Nitromethane and Liquid TNT', *Los Alamos Rep.* LA 3297 (1965).

32 EVANS, M. W., ABLOW, C. M., 'Theories of Detonation', *Chem. Rev.* **61**, 129–178 (1961).

33 KLINE, S. J., SHAPIRO, A. H., 'One-Dimensional, Steady Gas Dynamics for an Arbitrary Fluid', Publ. Sci. Techn. Ministère de l'Air, *Mémoires sur la Mécanique des Fluides* 171 (1954).

34 OSWATITSCH, K., 'Kondensationsstöße in Laval-Düsen', *Z.V.D.I.* **86** (1942) 702.

35 REED, S. G. (Jr.), 'Some Examples of Weak Detonations', *J. Chem. Phys.* **20**, 1823–24 (1952).

36 HEYBEY, W. H., REED, S. G. (Jr.), 'Weak Detonations and Condensation Shocks', *J. Appl. Phys.* **26**, 969–974 (1955).

37 STEPANCHUK, V. F., SALTANOV, G. A., 'Calculation of Oblique Condensation Shock Waves', *High Temp. Phys.* **3**, 550–559 (1965), (Trsl. from Russian).

Chapter IV

38 DAMKÖHLER, G., 'Einflüsse der Strömung, Diffusion und des Wärmeüberganges auf die Leistung von Reaktionsöfen', *Z. Elektrochem.* **42**, 846–862 (1936).

39 TOWNEND, L.H., 'An Analysis of Oblique and Normal Detonation Waves', Royal Aircraft Establishment, *Techn. Rep.* TR 66 081 (1966).

40 VON KÁRMÁN, TH., 'Aerothermodynamics and Combustion Theory', *L'Aerotecnica* **33**, 80 (1953).

41 PUKHNACHEV, V.V., 'The Stability of Chapman–Jouguet Detonations', *Sovj. Phys. Dokl.* **8**, 338–340 (1963).

42 ZIEREP, J., 'Über den Einfluß der Wärmezufuhr bei Hyperschallströmungen', *Acta Mech.* **2**, 217–230 (1966).

43 ADAMSON, T.C. (Jr.), MORRISON, R.B., 'On the classification of normal detonation waves', *Jet Propulsion* **25**, 400–403 (1955).

44 RICHTER, H., 'Stoß- und Detonationsfronten bei Abelscher Zustandsgleichung', *DEFA/LRBA-Rapport* 31/47 (1947) (ISL).

45 HIRSCHFELDER, J.O., CURTISS, C.F., BIRD, R.B., *Molecular Theory of Gases and Liquids*, John Wiley, New York; Chapman & Hall, London (1954).

46 SEMJONOW, N.N., *Einige Probleme der chemischen Kinetik und Reaktionsfähigkeit*, Akademie-Verlag, Berlin (1961).

47 DANIELS, F., *Outlines of Physical Chemistry*, John Wiley, Chapman & Hall, New York, London (1948).

Chapter V

48 BARTLMÄ, F., Instationäre Strömungsvorgänge bei Überschreiten der kritischen Wärmezufuhr. *Z. f. Flugwissensch.* **11**, 160–168 (1963).

49 STREHLOW, R.A., *Fundamentals of Combustion*, Intern. Textbook Co., Scranton, Pa. (1968).

50 VON NEUMANN, J., 'Theory of Detonation Waves', *OSRD Report* 549 (1942).

51 BRINKLEY, S.R. (Jr.), KIRKWOOD, J.G., 'On the condition of stability of the plane detonation wave', *III. Sympos. (Intern.) Comb. Flame a. Expl.*, Williams & Wilkins, 586–590 (1949).

52a SHCHELKIN, K.I., TROSHIN, YA.K., *Gasdynamics of Combustion*. Moscow (1963); Transl.: Mono Book Corp., Baltimore (1965).

52b SOLOUKHIN, R.I., *Shock Waves and Detonations in Gases*. Moscow (1963); Transl.: Mono Book Corp., Baltim. (1966).

52c SOLOUKHIN, R.I., *Non-stationary phenomena in gaseous detonation*. (Review Article.) Ref. 136, 799–807 (1969).

53 BERGER, J., VIARD, J., *Physique des explosifs solides*, Dunod, Paris (1962).

54 KISTIAKOWSKY, G.B., KYDD, J.P., 'Gaseous detonations IX. A study of the reaction zone by gas density measurements', *J. Chem. Phys.* **25**, 824–835 (1956).

55 LEWIS, B., VON ELBE, G., *Combustion, Flames and Explosion of Gases*, Academic Press Inc., New York (1951).

56 SOLOUKHIN, R.I., 'Multiheaded structure of gaseous detonation', *Comb. Flame* **10** 51–58 (1966).

57 BRINKLEY, S.R. (Jr.), RICHARDSON, J.M., 'On the structure of plane detonation waves with finite reaction velocity', *IV. Sympos. Combust.*, Williams & Wilkins, 450–457 (1953).

58 WOOD, W.W., KIRKWOOD, J.G., 'On the Existence of Steady-State Detonations Supported by a Single Chemical Reaction', *J. Chem. Phys.* **25**, 1276–1277 (1956).

59 Wood, W.W., Salsburg, Z.W., 'Analysis of Steady-State Supported One-Dimensional Detonations and Shocks', *Phys. Fluids* **3**, 549–566 (1960).

60 Erpenbeck, J.J., 'Steady Detonations in Idealized Two-Reaction Systems', *Phys. Fluids* **7**, 1424–1432 (1964); *Phys. Fluids* **4**, 481–492 (1961).

61 Fickett, W., Wood, W.W., 'Flow Calculations for Pulsating One-Dimensional Detonations', *Phys. Fluids* **9**, 903–916 (1966).

62 Davis, W.C., Craig, B.G., Ramsay, J.B., 'Failure of the Chapman-Jouguet Theory for Liquid and Solid Explosives', *Phys. Fluids* **8**, 2169–2182 (1965).

63 Chéret, R., 'L'onde de détonation idéale et la condition de Chapman–Jouguet', *Ann. Phys.* **1**, 169–179 (1966).

64 Eisen, C.L., Gross, R.A., Rivlin, T.J., 'Theoretical Calculations in Gaseous Detonation', *Combust. Flame* **4**, 137–147 (1960).

65 Wecken, F., 'Über eine Methode zur Bestimmung des Detonationsdruckes kondensierter Sprengstoffe', *LRSL-Note Technique* **4a/59** (1959) (ISL).

66 Wood, W.W., Fickett, W., 'Investigation of the Chapman–Jouguet Hypothesis by the "Inverse Method"', *Phys. Fluids* **6**, 648–652 (1963).

67 Petrone, F.J., 'Validity of the Classical Detonation Wave Structure for Condensed Explosives', *Phys. Fluids* **11**, 1473–1478 (1968); see also Ref. [105].

Chapter VI

68 Mader, C.L., 'Detonation Properties of Condensed Explosives Computed Using the Becker–Kistiakowsky–Wilson Equation of State', *Los Alamos Scientific Labor. Rep.* **LA–2900** (1963).

69 Wecken, F., 'Stoßwellen in einem van-der-Waals-Gas', *DEFA/LRBA-Rapport* **4/48** (1948) (ISL).

70 Deal, W.E., 'Measurement of the Reflected-Shock Hugoniot and Isentrope for Explosive Reaction Products', *Phys. Fluids* **1**, 523–527 (1958).

71 Fickett, W., Wood, W.W., 'A Detonation-Product Equation of State Obtained from Hydrodynamic Data', *Phys. Fluids* **1**, 528–534 (1958).

72 Berger, J., Favier, J., Nault, Y., 'Détermination des caractéristiques de détonation des explosifs solides', *Annales de Physique* **5**, 771–803 and 1143–1176 (1960).

73 Penner, S.S., *Chemistry Problems in Jet Propulsion*, Pergamon Press, London (1957).

74 Schmidt, A., 'Tabellen thermochemischer Daten von Pulvern, Explosivstoffen und Bestandteilen explosiver Gemische', *LRSL Rapport* **3/53** (1953) (ISL).

75 Filler, W.S., 'Post-detonation pressure and thermal studies of solid high explosives in a closed chamber', *Sixth Symposium on Combustion*, p. 648–657. Reinhold, New York (1957).

76 Ulich, H., Jost, W., *Kurzes Lehrbuch der physikalischen Chemie. 16. Aufl.*, Darmstadt (1966).

77 *Kinetics, Equilibria and Performance of High Temperature Systems*, Proceedings of Conference 1959, Butterworths, London (1960).

78 Vidart, A., 'Calcul des caractéristiques de détonation', *Mém. des Poudres* **42**, 83–144 (1960).

79 Chevance, R., Contribution au calcul des équilibres chimiques en phase gazeuse homogène', *Dir. Techn. des Armements Terrestres*, Paris, 9994/CEPA-RA (1966).

80 Brinkley, S.R. (Jr.), 'Computational methods in combustion calculation', *Combustion processes = High Speed Aerodynamics*, Vol. 2, p. 64–98. Princeton (1956).

Chapter VII

81 *'Hütte' – des Ingenieurs Taschenbuch,* Bd. 1, Berlin 1942 (27. Aufl.) und 1955 (28. Aufl.) Artikel 'Ähnlichkeitsmechanik und Modellwissenschaft' von M. Weber, S. 435–445 (1942); 'Ähnlichkeitsmechanik und Modelltechnik' von A. Betz, S. 744–752 (1955).

82a GUKHMAN, A. A., *Introduction to the Theory of Similarity*, Academic Press, New York–London (1965).

82b KATTANEK, S., GRÖGER, R., BODE, C., *Ähnlichkeitstheorie*, VEB Deutscher Verlag für Grundstoffindustrie, Leipzig 1967.

83 TAYLOR, G. I., The dynamics of the combustion products behind plane and spherical detonation fronts in explosives, *Proc. Roy. Soc.* **A 200,** 235–247 (1950).

84 WECKEN, F., 'Ähnlichkeitsgesetze bei Explosionen', *LRSL* **5a/58** (1958).
Les lois de similitude dans les explosions à symétrie sphérique. *Mém. Art. franç.* **35**, 2e fasc., 437–459 (1961).

85 WECKEN, F., 'Pseudostationäre Strömungsvorgänge', *LRSL* **26/46** (1946).

86 SEDOV, L. I., *Similarity and Dimensional Methods in Mechanics*, Infosearch Ltd., London (1959).

87 SCHARDIN, H., 'Ein Beispiel zur Verwendung des Stoßwellenrohres für Probleme der instationären Gasdynamik', *ZAMP* **9b**, 606–621 (1958).

88 WECKEN, F., MÜCKE, L., 'Die Detonation einer kugelförmigen Sprengladung', Teil 1, *LRSL* **8/50** (1950).

Chapter VIII

89 BARTLMÄ, F., Schiefe Reaktionsfronten. *Deutsche Luft- und Raumfahrt, Forschungsber.* 67–73 = *DVL-Ber.* 684 (1967).

90 ROSS, F. W., 'The propagation in a compressible fluid of finite oblique disturbances with energy exchange and change of state', *J. appl. Phys.* **22**, 1414–1421 (1951).

91 LARISCH, E., 'Interactions of detonation waves', *J. Fluid Mech.* **6**, 392–400 (1959).

92 SIESTRUNCK, R., FABRI, J., LE GRIVES, E., 'Some properties of stationary detonation waves', *Fourth Symposium on Combustion, M.I.T.*, 1952, Williams and Wilkins, p. 498–501 (1953).

93 CHINITZ, W., BOHRER, L. C., FOREMAN, K. M., 'Properties of oblique detonation waves', *AFOSR-TN* 59–462, ASTIA No. AD 215 267 (1959).

94 GROSS, R. A., 'Oblique detonation waves', *AIAA J.* **1**, 1225–1227 (1963).

95 SICHEL, M., 'A hydrodynamic theory for the interaction of a gaseous detonation with a compressible boundary', *Univ. Michigan Dpt. of Aeron. and Astron. Engg., Techn. Rep.* (1965).

96 GROSS, R. A., 'A study of combustion in supersonic flow', *Research appl. in Industry* **12**, 381–389 (1959).

97 WECKEN, F., 'Das Verhalten von Stoßwellen an Materiegrenzen', *Lab. Rech. Techn.*, St-Louis (France), Rapp. 29/47 (1947).

98 PACK, D. C., 'The reflection and diffraction of shock waves', *J. Fluid Mech.* **18**, 549–557 (1964).

99 DUNNE, B. B., 'Mach reflection of detonation waves in condensed high explosives', *Phys. Fluids* **4**, 918–924, 1565–1566 (1961); **7**, 1707–1712 (1964).

100 HILL, R., PACK, D. C., 'An investigation, by the method of characteristics, of the

lateral expansion of the gases behind a detonating slab of explosive', *Proc. Roy. Soc. (A)* **191**, 524–541 (1947).

101 WECKEN, F., MÜCKE, L., 'Die Detonation einer kugelförmigen Sprengladung. Teil 2', *LRSL* **1/53** (1953).

102 'Les ondes de détonation', *Colloque CNRS*, Gif-sur-Yvette 1961, Centre National de Rech. Scient., Paris (1962).

103 *Fourth Symposium on Detonation*, US-NOL, White Oak 1965, ONR, Rep. ACR-126 (1966).

104 OPPENHEIM, A. K., 'Gasdynamics of Explosions', Chapter 13 in *Jet, Rocket, Nuclear, Ion and Electric Propulsion, Theory and Design*, Editor, W.H.T. Loh. Springer-Verlag, New York, Inc. (1968).

105 *High Dynamic Pressure*, IUTAM Symposium Paris 1967, Dunod, Paris and Gordon and Breach, New York (1968).

106 SCHALL, R., *Dynamische Hochdruckphysik. Beiträge zur Ballistik und Technischen Physik*, Gedenkschrift für Hubert Schardin, Beiheft Nr. 7 der Wehrtechnischen Monatshefte, 128–143 (1967).

Chapter IX

107 SCHALL, R., 'Detonationsphysik', *Kurzzeitphysik*, editors K. Vollrath and G. Thomer, Springer Wien–New York, 849–907 (1967).

108 EYRING, H., POWELL, R. E., DUFFEY, G. H., PARLIN, R. B., 'The stability of detonation', *Chem. Review* **45**, 69–181 (1949).

109 JONES, H., 'A theory of the dependence of the rate of detonation of solid explosives on the diameter of the charge', *Proc. Roy. Soc.* **A 189**, 415–426 (1947).

110 WECKEN, F., 'Non-ideal detonation with constant lateral expansion', [103], 107–116 (1966); *Technische Mitteilung ISL* **T 19/65**.

111 EVANS, M. W., 'Detonation sensitivity and failure diameter in homogeneous condensed materials', *J. Chem. Phys.* **36**, 193–200 (1962).

112 WOOD, W. W., KIRKWOOD, J. G., 'Diameter effect in condensed explosives. The relation between velocity and radius of curvature of the detonation wave', *J. Chem. Phys.* **22**, 1920–1924 (1954).

113 KEGLER, W., 'Sensibilisierung von Trinitrotoluol durch Nitrozellulose und andere Zusätze', *Technische Mitteilung ISL* **T 7/63** (1963).

114 COOK, M. A., *The science of high explosives*, Reinhold, New York (1958).

115 FAY, J. A., 'Two-dimensional gaseous detonations: velocity deficit', *Phys. Fluids* **2**, 283–289 (1959).

116 STRUTH, W., BEHRENS, H., WECKEN, F., 'Untersuchung chemischer Reaktionen bei Einschuß in reagierende Gase oder Gasgemische mit hohen Geschwindigkeiten', *Technische Mitteilung ISL* **T 9/63** (1963).

117 SAMOZVANTSEV, M. P., 'The stabilisation of detonation waves by means of bluff bodies', *Prikl. Mat. tekh. Fiz.* No. **4**, 126–129 (1964); *RAE*, Farnborough, Transl. No. **1088** (1965).

118a SOMMERS, W. P., 'Gaseous detonation-wave interactions with non-rigid boundaries', *ARS J.* **31**, 1780–1782 (1961).

118b SOMMERS, W. P., MORRISON, R. B., 'Simulation of condensed-explosive detonation phenomena with gases', *Phys. Fluids* **5**, 241–248 (1962).

119a DABORA, E.K., 'The influence of a compressible boundary on the propagation of gaseous detonations', *Univ. Michigan, Coll. of Eng., Dept. of Aeron. and Astronaut. Eng., Aircr. Propuls. Lab., Techn. Rept.* (1963).

119b DABORA, E.K., NICHOLLS, J.A., MORRISON, R.B., 'The influence of a compressible boundary on the propagation of gaseous detonations', *10th Sympos. on Combustion. The Combust. Inst.*, Pittsburgh, Pa., 817–830 (1965).

120a SICHEL, M., 'A hydrodynamic theory for the interaction of a gaseous detonation with a compressible boundary', *Univ. Michigan, Coll. Eng., Techn. Rep.* (1965).

120b 'A hydrodynamic theory for the propagation of gaseous detonations through charges of finite width', *AIAA J.* **4**, 264–272 (1966).

121 KERKAM, B.F., DABORA, E.K., 'Effect of water vapor on H_2-O_2 detonations', *AIAA J.* **4**, 1101–1102 (1966).

122 MAČEK, A., 'Transition from deflagration to detonation in cast explosives', *J. Chem. Phys.* **31**, 162–167 (1959).

123 ZOVKO, C.T., MAČEK, A., *A computational treatment of the transition from deflagration to detonation in solids*, [130] II, 606–634 (1960).

124 HUBBARD, H.W., JOHNSON, M.H., 'Initiation of detonations', *J. appl. Phys.* **30**, 765–769 (1959).

125 BERNIER, H., 'Contribution à l'étude de la génération de la détonation provoquée par impact sur un explosif', *Centre d'Etudes Nucléaires de Saclay, Rapp. CEA-R* **2497** (1964); Engl. transl. **AD 628 984.**

126 BERNIER, H., LEZAUD, J.M., CHEVALIER, Y., *Etude expérimentale et théorique de la génération de la détonation par onde de choc dans un explosif granulaire*, [105], 51–57 (1968).

127 MADER, C.L., 'Shock and hot spot initiation of homogeneous explosives', *Phys. Fluids* **6**, 375–381 (1963).

128 MADER, C.L., 'Initiation of detonation by the interaction of shocks with density discontinuities', *Phys. Fluids* **8**, 1811–1816 (1965); *Fourth Symposium on Detonation*, Preprints, **C 38–C 44** (1965).

129 ERPENBECK, J.J., 'Steady-state analysis of quasi one-dimensional reactive flow', *Preprint from LASL, LA-DC-***9048** (1967).

130 *Third Symposium on Detonation.* I, II. ONR Symposium Report ACR-**52** (1960).

131 BOWDEN, F.P., YOFFE, A.D., *Initiation and growth of explosions in liquids and solids*, Cambr. Univ. Press, London (1952).

132 BOWDEN, F.P., YOFFE, A.D., *Fast reactions in solids*, Butterworths, London (1958).

133 'A discussion of the initiation and growth of explosion in solids', Under the leadership of F.P.BOWDEN. *Proc. Roy. Soc. A* **246**, 145–297 (1958).

134 SKIDMORE, I.C., *Computer studies of shock initiation in homogeneous and heterogeneous explosives*, [105], 43–49 (1968).

135a WATSON, R.W., SUMMERS, C.R., GIBSON, F.C., VON DOLAH, R.W., *Detonations in liquid explosives—the low-velocity regime*, [103], 117–125 (1965).

135b WATSON, R.W., *The structure of low-velocity detonation waves*, [136], 723–729 (1969).

136 *12th Symposium on Combustion*, The Combustion Institute, Pittsburgh Pa. (1969).

137 PRICE, D., Contrasting patterns in the behavior of high explosives. *11th Symposium on Combustion* (Berkeley 1966), 693–702, The Combustion Institute, Pittsburgh, Pa. (1967).

B. Selected Literature on Detonics, Shock Waves and High Pressure Physics

a) Symposia on Detonation in the USA

Sponsored by Office of Naval Research, Washington, D.C., and Naval Ordnance Laboratory, Silver Spring, Md.

(1st) Conference on the Chemistry and Physics of Detonation (1951). Conference classified but proceedings later declassified. AD 127020.

Second ONR Symposium on Detonation (1955). Two volumes: Unclassified volume PB 128369, Confidential volume AD 52145.

Third Symposium on Detonation, Princeton University (1960). ONR Sympos. Rep. ACR-52; vol. 1, 2 unclassified, PB 181172/73, vol. 3 confidential, AD 322061. See Ref. [130].

Fourth Symposium (International) on Detonation, NOL, Silver Spring (1965). See Ref. [103].

Fifth Symposium on Detonation, Pasadena, Cal. (1970). To be published.

b) International Symposia on Combustion

Organized by The Combustion Institute, Pittsburgh, Pa., held in even years and published in odd years; see Ref. [51], [57], [92], [75], [119], [136].

13th Symposium, Univ. Utah, Salt Lake City (1970). To be published.

c) International Colloquia on Gasdynamics of Explosions

First Colloquium, Univ. Brussels, Belgium (1967). Proceedings: Astronaut. Acta **14**, Special Issue no. 5 (1969).

Second Colloquium, Novosibirsk (1969). Astronaut. Acta **15**, 253–661 (1970).

d) International Shock Tube Symposia

5th, NOL (1965). Proceedings published by NOL (1966).

6th, Ernst-Mach-Institut Freiburg, W.Germany (1967). Proceedings: Phys. Fluids **12**, Special Issue no. 5 II (1969).

7th, Univ. Toronto (1969). Univ. Toronto Press (1970).

e) Other Symposia on Related Topics

General Discussion on the Physical Chemistry of Processes at High Pressures, Chemistry Dept., Univ. Glasgow (1956). Discussions of the Faraday Soc. no. 22 (1956).

International Conference on Sensitivity and Hazards of Explosives. Sponsored by Expl. Res. Devel. Establ., Waltham Abbey, Essex, England. London (1963).

See also Ref. [102], [105], [133].

f) Monographs and Survey Articles

Advances in High Pressure Research. Editor: R.S. Bradley. Academic Press, vol. 1 (1966), vol. 2, 3 (1969).

ANDREJEV, K.K., Thermische Zersetzung und Verbrennungsvorgänge bei Explosivstoffen. Mannheim (1964).

ANDREJEV, K.K., BELJAJEV, A.F., Theorie der Explosivstoffe. 3 vols. In Russian, Moscow (1960). German translation, Svenska Nationalkommittén för Mekanik, Stockholm (1964).

ASCANI, D.C., The Literature on Explosives. Advances in Chemistry series no. 76. Amer. Chem. Soc., Washington, D.C. (1968).

BAUM, F.A., STANYUKOVICH, K.P., SHEKHTER, B.I., Physics of an Explosion. In Russian, Moscow (1959). Translation, Res. Inform. Serv., New York (1963).

BRADLEY, J.N., Shock Waves in Chemistry and Physics, London and New York (1962).

BRIDGMAN, P.W., The Physics of High Pressure, London (1949).

BRIDGMAN, P.W., Collected Experimental Papers, Harvard Univ. Press (1964).

COLE, R.H., Underwater Explosions, Princeton Univ. Press (1948).

High Pressure Physics and Chemistry. Editor: R.S.Bradley. Two vols. Academic Press (1963).

HIRSCHFELDER, J.O., and others, Theory of Detonations. J. Chem. Phys. **28**, 1130–1151 (1958), **30**, 470–492 (1959).

JACOBS, S.J., Recent Advances in Condensed Media Detonations. ARS J. **30**, 151–158 (1960).

JOHANSSON, C.H., PERSSON, P.A., Detonics of High Explosives, Academic Press (1970).

Modern Very High Pressure Techniques. Editor: R.H.Wentorf, Jr. Butterworths (1962).

OPPENHEIM, A.K., MANSON, N., WAGNER, H.G., Recent Progress in Detonation Research. AIAA J. **1**, 2243–2252 (1963).

Physics of High Pressures and the Condensed Phase. Editor: A. van Itterbeek. North-Holland, Amsterdam (1965).

RINEHART, J.S., PEARSON, J., Explosive Working Metals, Pergamon Press (1963).

Solids Under Pressure. Editors: Paul, W., Warschauer, D.M. McGraw-Hill (1963).

STANYUKOVICH, K.P., Unsteady Motion of Continuous Media, Pergamon Press (1960).

TAYLOR, J., Detonation in Condensed Explosives, Clarendon Press (1952).

WILLIAMS, F.A., Combustion Theory, Addison-Wesley (1965).

Index

U.C.W. ABERYSTWYTH
LIBRARY